SUPERFUND'S FUTURE

WHAT WILL IT COST?

A Report to Congress

Katherine N. Probst and David M. Konisky

with Robert Hersh, Michael B. Batz, and
Katherine D. Walker

RESOURCES FOR THE FUTURE
Washington, DC

©2001 Resources for the Fu

Printed in the United States of America

An RFF Press book
Published by Resources for the Future
1616 P Street, NW, Washington, DC 20036–1400
www.rff.org

Library of Congress Cataloging-in-Publication Data

Probst, Katherine N.
    Superfund's future : what will it cost? : a report to Congress / Katherine N. Probst and David M. Konisky ; with Robert Hersh, Michael B. Batz, and Katherine D. Walker.
        p.   cm.
    Includes bibliographical references and index.
    ISBN 1–891853–39–2 (pbk.)
    1. United States. Comprehensive Environmental Response, Compensation, and Liability Act of 1980. 2. Hazardous waste sites—Cleaning—United States—Finance. 3. Pollution—Economic aspects—United States. I. Konisky, David M. II. Hersh, Robert. III. Title
    HC110.W3 P76 2001
    363.738′4—dc21   2001031741

f   e   d   c   b   a

The paper in this book meets the guidelines for permanence and durability of the Committee on Production Guidelines for Book Longevity of the Council on Library Resources.

Cover photograph by David Noton. The cover, interior, and figures for this book were designed by Debra Naylor Design. The text was composed by Betsy Kulamer. It was copyedited by Sally Atwater.

Funds for the printing of this book were provided entirely by Resources for the Future.

The findings, interpretations, and conclusions offered in this publication are those of the authors and are not necessarily those of Resources for the Future, its directors, or its officers.

ISBN 1–891853–39–2 (paper)

About Resources for the Future and RFF Press

Founded in 1952, **Resources for the Future** (RFF) contributes to environmental and natural resource policymaking worldwide by performing independent social science research.

RFF pioneered the application of economics as a tool to develop more effective policy about the use and conservation of natural resources. Its scholars continue to employ social science methods to analyze critical issues concerning pollution control, energy policy, land and water use, hazardous waste, climate change, biodiversity, and the environmental challenges of developing countries.

RFF Press supports the mission of RFF by publishing book-length works that present a broad range of approaches to the study of natural resources and the environment. Its authors and editors include RFF staff, researchers from the larger academic and policy communities, and journalists. Audiences for RFF publications include all of the participants in the policymaking process—scholars, the media, advocacy groups, nongovernment organizations, professionals in business and government, and the general public.

Contents

List of Tables
and Figures

List of Tables

List of Figures

Acknowledgements

Our first thanks go to the congressional requesters of this study, Carrie Apostolou, formerly with the Senate Appropriations Committee, and Frank Cushing, of the House Appropriations Committee. We were extraordinarily lucky to have congressional clients who were extremely supportive, proffered excellent advice, and allowed us to conduct our work unfettered.

The research in this book could not have been undertaken without the personal and unwavering support of Tim Fields, the former Assistant Administrator of the Office of Solid Waste and Emergency Response (OSWER) at the U.S. Environmental Protection Agency (EPA). When the congressional request for this report was first announced, Tim made clear his commitment that Resources for the Future (RFF) would enjoy the full cooperation of the Agency and that we would be able to conduct this research in an independent manner. The research in this book was funded under a contract with OSWER (EPA Contract 68-W-00-079).

Three Deputy Assistant Administrators for OSWER were also instrumental in assuring that we got the information we needed from EPA: former Deputy Assistant Administrator Cliff Rothenstein, who oversaw EPA's efforts in the early stages of the project; Michael Shapiro, who has served as acting Assistant Administrator since Tim's departure in January 2001 and has continued in this capacity—as before—to assure our independence in this effort; and Steve Luftig, who has provided useful information about the history of the Superfund program. Mike Ryan, EPA's Deputy Chief Financial Officer, was also extremely supportive of our efforts.

Equally important to the success of our efforts was the excellent work of one of Tim Fields' budget staff members, Tim Fontaine, who was our primary contact at the Agency. Tim Fontaine had the thankless job of responding to our unending requests for data, information, and meetings. He did this with grace, professionalism, and the curiosity of a true policy analyst. Without the personal commitment of Tim Fontaine and Bruce Pumphrey, of the enforcement program, our task would have been much more difficult. We are extremely grateful to them both for their sincere and unrelenting dedication to improving the Superfund program.

A large group of other EPA staff members provided us with information for our work over the course of the project. These included three regional staff brought to Washington to help the Agency manage the study from EPA's end: Don Williams (Region 6), Kathi Moore (Region 9), and Glenn Kistner (Region 9), as well as a core group of people in EPA headquarters: Erin Conley, Mike Cullen, Dave Evans, Russ Harmon, Randy Hippen, Tracy Hopkins, Art Johnson, Dan Malloy, and Alan Youkeles. Many other staff in EPA headquarters, far too numerous to name here, also contributed to our efforts; we sincerely appreciate their assistance.

In addition to staff in EPA headquarters, we were fortunate to work closely with the Superfund division directors in all 10 EPA regional offices. These individuals—Richard Caspe, Max Dodson, Abe Ferdas, Michael Gearhard, Richard Green, Myron Knudson, Patricia Meaney, Bill Muno, Michael Sanderson, and Keith Takata—added an important perspective that enhanced our understanding of the Superfund program. Each of them generously gave their time to participate in our interviews about the number and character of future National Priorities List (NPL) sites, and they supported our request to gather site-specific information about 112 of the most expensive sites on the NPL. We also thank the regional removal program managers who responded to our questions about the removal program, as well as the regional coordinators for the RFF study and other regional staff who processed myriad requests for information and clarifications.

We also owe our sincere thanks to the directors of the nine state superfund programs who participated in our survey regarding state programs and future NPL sites. These state officials—Gary Behrns, Barbara Coler, Dan DiDomenico, Jim Feeley, Thomas Fidler, Cindy Hafner, Jay Naparstek, Michael O'Toole, and Ed Putnam—graciously gave of their time to participate in our state interviews.

An additional group of almost 30 people from congressional committee staffs, the U.S. Department of Justice (DOJ), state agencies and associations, the private sector, and the environmental community spent hours in meetings at RFF, critiquing first our approach and then the preliminary results of our cost model; finally, they—along with many EPA and state agency staff—reviewed a confidential draft of this book to help us identify and correct errors.

Two people deserve particular recognition for going above and beyond the call of duty in this regard: Perry Beider of the Congressional Budget Office and Joe Hezir of the EOP Group. Both of these Superfund experts helped us enormously. Other individuals who critiqued our work and whom we wish to thank are congressional staffers Jim Barnette, Susan Bodine, David Conover, Dick Frandsen, Ken Kopocis, and Barbara Rogers, and former staffer Kirstin Rohrer; Bob Bruffy, John Cruden, Bruce Gelber, and Steve Gold of DOJ, as well as Lois Schiffer, the former Assistant Attorney General for Environment and Natural Resources; state program managers Jack Butler, Mark Giesfeldt, Gary King, Craig Olewiler, Mark Schmitt, George Schuman; industry representatives Keith Belton, Kerry Kelly, Dell Perelman, Mike Toohey, and John Ulrich; environmental group representatives Karen Florini and Grant Cope; state association representatives Tom Kennedy, Diane Shea, and Tim Titus; and Superfund experts Jan Acton and Mike Tilchin.

All of the people acknowledged here gave freely of their time to help make this report a better product, and we are extremely grateful. However, nothing in this report—including our findings and recommendations—should be assumed to have their agreement or support, and any remaining errors are our own.

The research, production, and distribution of this report would not have been possible without the support of many people at RFF, as well as several RFF contractors. These include staff in our communications office, Felicia Day, Jonathan Halperin, and Scott Hase; staff of the RFF Press, Gina Armento and Don Reisman; RFF researcher Sandy Hoffmann; as well as former RFF staff Matt Cannon, Dan Quinn, Ken Rubotzky, and Molla Sarros. We also want to thank our copyeditor, Sally Atwater; layout expert, Betsy Kulamer; graphics guru, Debra Naylor; and RFF staff who came to our aid at various points during the project—Aracely Alicea, Aris Awang, and María Reff.

Finally, we want to thank our contributors to this report, Bob Hersh, Michael Batz, and Katy Walker, all of whom were instrumental to this project. Bob was largely responsible for designing the interview guides for our surveys of the EPA regional Superfund division directors and state superfund managers and is the lead author of Chapter 5. Michael was our computer maven; he not only built our cost model but also checked, double-checked, and then checked again all our numbers as well as produced all our tables and figures. Katy delved into the removal program and is the lead author of Chapter 3.

KATHERINE N. PROBST AND DAVID M. KONISKY
Resources for the Future

About the Authors

Katherine N. Probst is a senior fellow at Resources for the Future (RFF). During the past 20 years, she has directed major studies of the Superfund program. She is the lead author of RFF's previous book on Superfund, *Footing the Bill for Superfund Cleanups: Who Pays and How?* She is also the lead author of many RFF reports, including: *Cleaning Up the Weapons Complex: Does Anybody Care?*, *The Evolution of Hazardous Waste Programs: Lessons from Eight Countries,* and *Long-Term Stewardship and the Nuclear Weapons Complex: The Challenge Ahead.* She was the project director of RFF's evaluation of the role of land use and Superfund cleanups.

David M. Konisky is a research associate at RFF and will begin doctoral studies in the department of political science at the Massachusetts Institute of Technology in the fall of 2001. His recent research has focused on the Superfund program and public participation in environmental decisionmaking.

Robert Hersh is a fellow at RFF. His current research interests include brownfields redevelopment, land use, and social theories of risk. Since 1992, much of his work at RFF has sought to answer the question of how diverse interests at the state and local level—residents, government officials, industry, public interest groups—can reach agreement to better manage environmental and health risks. He is lead author of the RFF report, *Linking Land Use and Superfund Cleanups: Uncharted Territory,* and has been published in numerous journals.

Michael B. Batz is a research assistant at RFF. He received a master of science degree in electrical and computer engineering from Carnegie Mellon University in 1998.

Katherine D. Walker is an independent consultant on environmental health issues. She began working with the Superfund program in 1982, when she was in charge of conducting health risk assessments at hazardous waste sites for major U.S. engineering firms. More recently, she was director of the Superfund Reform Project at the Harvard Center for Risk Analysis. She is a contributor to the RFF book, *Analyzing Superfund: Economics, Science, and Law.*

Executive Summary

The debate about whether and how to reauthorize the Comprehensive Environmental Response, Compensation, and Liability Act of 1980 (CERCLA)—better known as Superfund—has continued relatively unabated since 1986, when the law was last the subject of a major reauthorization. A central question now facing Congress is whether there is enough money in the Hazardous Substance Superfund (Trust Fund) to continue to pay for the program. If not, funds for the program will need to come from general revenues, unless the now-expired taxes that stocked the Trust Fund are reimposed. As a result, the question of whether, and when, the Superfund program will "ramp down" is a pressing concern.

To shed some light on this question, Congress asked Resources for the Future (RFF) to estimate the future cost of the Superfund program. More specifically, as part of the conference report that accompanied the FY 2000 Departments of Veterans Affairs and Housing and Urban Development, and Independent Agencies appropriations bill, RFF was asked to conduct an independent study to estimate how much money will be needed by the U.S. Environmental Protection Agency (EPA) to implement the Superfund program from FY 2000 through FY 2009. Congress identified six elements for the study:

1. the remaining cost of cleaning up the sites on EPA's National Priorities List (NPL) at the end of FY 1999—the sites at which the Agency can pay for long-term cleanups with monies from the Trust Fund;
2. the cost of cleaning up sites added to the NPL from FY 2000 through FY 2009;
3. the cost of conducting emergency response and removal actions, which are generally shorter in duration and less expensive than a typical NPL cleanup;

4. the cost of performing five-year reviews, which are required at most NPL sites to make sure that remedies remain protective over time;
5. the cost of implementing long-term response actions (a specific type of operation and maintenance activity) at NPL sites; and finally,
6. the cost of administering the Superfund program, including policy development, program support and management, rent, contract and information management, and accounting systems, as well as a number of Superfund-related programs at EPA and other federal agencies.

The total cost to EPA of implementing the Superfund program in FY 1999 added up to $1.54 billion.*

The congressional language requesting this report specifically excludes from our estimates the cost of cleaning up sites owned and operated by federal agencies (e.g., the Department of Energy and the Department of Defense), as well as the cost to EPA of overseeing these cleanups. The cost to EPA of carrying out the brownfields program is also excluded, as are Superfund costs to potentially responsible parties and state environmental agencies.

Several other aspects of the task assigned to RFF are important to note. First, our estimates represent the cost of implementing the program under current law—in other words, assuming no change in the law's existing liability or cleanup standards or in current EPA policies. Second, our estimates are not constrained by past funding levels. That is, our approach is to model the future work of the Superfund program and, based on that, derive estimates of the funds needed. To estimate future costs, we focus on EPA *expenditures*. Our estimates of the funds needed to implement the Superfund program thus reflect the amount of money EPA will actually spend each year. These estimates will need to be translated into obligations for the appropriations process.

It is important, at the outset, to acknowledge the uncertainty of the estimates presented in this report—an issue anticipated in the conference report language. Our main limitations are the quality of the data available and, of course, our ability to predict the future. Our analysis relies primarily on existing EPA information, although we collected new information in several key areas.

We address uncertainty by developing estimates under three scenarios—a base case, a low case, and a high case. For sites on the NPL as of the end of FY 1999 (which we refer to as "current" NPL sites), we vary the cost of cleanup actions in each scenario. For sites added to the NPL from FY 2000 through

---

*This does not include the cost of implementing parts of the Superfund program specifically excluded from the RFF study.

FY 2009 ("future" NPL sites), the three scenarios vary in terms of the number of sites listed each year, the percentage of sites with total expected cleanup costs of $50 million or more (mega sites), and the share of cleanup actions paid for with Trust Fund monies (Fund-lead actions). Each of the scenarios leads to a different workload for the Superfund program, which also affects our estimate of other program costs.

We cannot overemphasize that our estimates are just that, estimates. It is likely that our estimates for the next few years are more reliable than for later years and that our estimates of the remaining costs of cleaning up current NPL sites are more reliable than our estimates for future sites. Given the full body of our research, analyses, and interviews, the base-case scenario represents our best estimate of the future cost of the Superfund program; the low and high cases provide a range of likely future costs.

In the course of our research, we identified issues that deserve closer attention from EPA, Congress, and others charged with implementing the Superfund program. We summarize our major findings and conclusions below and then conclude with recommendations for four issues we think need to be addressed to improve the Superfund program.

Findings and Conclusions

1. A ramp-down of the Superfund program is not imminent.

Under the base-case scenario, which represents our best estimate of the likely future cost of the Superfund program under current law and policies, EPA's need for Superfund monies will not decrease appreciably below FY 1999 expenditures of $1.54 billion until FY 2006, as shown in Figure ES-1. In the base case, total annual costs peak in FY 2003, driven principally by the cost of Fund-lead actions at a few mega sites, and then begin a steady but small decline each year. In FY 2009, the final year of our estimates, the total annual cost under the base-case scenario is $1.33 billion, which is 14% less than FY 1999 expenditures. If inflation is taken into account, our estimate of the total annual cost of the Superfund program in FY 2009 under the base-case scenario is $1.61 billion, which would represent a 5% increase over FY 1999 expenditures.

Even under the low-case scenario, annual Superfund costs do not fall dramatically. In the low case, the estimated total annual cost to EPA of implementing the Superfund program does not drop below $1.4 billion until FY 2005, and below $1.2 billion until FY 2009.

Figure ES-1. Estimated Total Annual Cost to EPA of Superfund Program: Three Scenarios, FY 2000–FY 2009 (1999$)

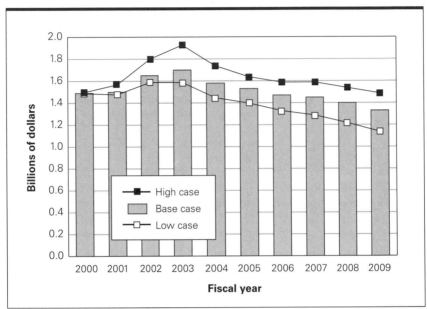

Overall, from FY 2000 through FY 2009, our estimates of total annual costs under the low-case scenario average 15% less than under the high-case scenario. The difference between the annual costs of these two scenarios is more pronounced, however, in the latter years of this period. By FY 2009, the low- and high-case cost estimates are $1.14 billion and $1.49 billion, respectively—a difference of $350 million, or 23% of FY 1999 expenditures. If the low estimate is correct, program expenditures would fall substantially during the last few years of the FY 2000 through 2009 period. However, if the high case proves correct, program expenditures in FY 2009 would be just 3% less than in FY 1999.

2. The total estimated cost to EPA of implementing the Superfund program from FY 2000 through FY 2009 ranges from $14 billion to $16.4 billion.

We developed six program elements to address the requirements in the congressional language requesting this report, while taking into account limitations in how EPA's expenditure data are organized. Our estimates of future costs are organized around these six elements:

1. *removal program:* emergency response and other short-term cleanup actions at NPL and non-NPL sites, including enforcement and community involvement activities, as well as national policy development, management, and response support activities;

2. *remedial program:* site studies and response actions conducted primarily but not exclusively at NPL sites, as well as other site-specific activities, such as enforcement, oversight, and community relations;

3. *site assessment:* activities undertaken to evaluate the severity of contamination at sites and to determine what, if any, response is needed;

4. *program staff, management, and support:* national policy development and program management for the remedial program, as well as for the overall Superfund program, including enforcement policy and management, support activities, grants to state programs, and a variety of other efforts;

5. *program administration:* rent, facilities operation and maintenance, equipment and supplies, and activities not charged specifically to sites that are carried out by regional staff and headquarters staff in EPA's Office of Administration and Resources Management and Office of the Chief Financial Officer; and

6. *other programs and agencies:* Superfund-related work of offices within EPA, such as the Technology Innovation Office and the Office of the Inspector General, and at other federal agencies, such as the National Institute for Environmental Health Sciences.

Table ES-1 presents our estimates for the six elements and the total 10-year cost of the Superfund program under each of the three scenarios. Under the base-case scenario, EPA will need $15.1 billion for FY 2000 through FY 2009 to implement the Superfund program. Under the low-case scenario, we estimate that the total cost over these 10 years will be approximately $14 billion, or about 7% less than in the base case. Under the high-case scenario, we estimate that the 10-year cost will be $16.4 billion, or about 9% more than in the base case. If we take into account likely future inflation, the range of total costs for FY 2000 through FY 2009 is $15.6 billion to $18.3 billion.

3. Fund-lead actions at current NPL sites will be the major driver of EPA cleanup costs from FY 2000 through FY 2009.

EPA has made considerable progress addressing sites on the current NPL. At the end of FY 1999, the Agency had designated just over half (52%) of all nonfederal NPL sites as "construction-complete" (this increased to 57% by the end of FY 2000). Yet there is still considerable cleanup work to be done at

Table ES-1. Estimated Total Cost to EPA of Superfund Program: Three Scenarios, FY 2000–FY 2009 (Billions of 1999$)

	Base Case	Low Case	High Case
Removal program	3.18	3.18	3.18
Remedial program	6.55	5.71	7.55
Site assessment	0.58	0.58	0.58
Program staff, management, and support	2.69	2.51	2.89
Program administration	1.10	1.02	1.20
Other programs and agencies	1.01	1.01	1.01
Total	**15.10**	**13.99**	**16.40**

Note: Totals may not add exactly because of rounding.

these sites. In fact, from FY 2000 through FY 2009, the estimated cost of Fund-lead actions at current NPL sites is much greater than that for actions at sites that will be added to the NPL.

Figure ES-2 shows the annual cost of Fund-lead actions at both current and future NPL under the base-case scenario. Future NPL sites do not account for the majority of the annual costs of Fund-lead actions until FY 2009, even in the high-case scenario. (In the low-case scenario, Fund-lead actions at current NPL sites still constitute the majority of costs in FY 2009.) In aggregate, from FY 2000 through FY 2009, the total cost of Fund-lead actions at all NPL sites under the base-case scenario is $5.28 billion, 84% of which will be spent at current NPL sites.

4. The number, type, and cost of future NPL sites are each difficult to predict.

No other component of our estimates is as fraught with uncertainty as the future composition of the NPL. Based on our analysis of recent listing trends and our interviews with all 10 EPA regional offices and with 9 state superfund program managers, we estimate that 23 to 49 sites will be added to the NPL each year from FY 2001 through FY 2009. (We use the actual number of final NPL listings in FY 2000, 36 sites, in all three of our future NPL scenarios.)

Most attempts to forecast the size of the NPL, and the associated costs, have focused on estimating the number of final listings. The total number of sites, however, is only one piece of the puzzle. To estimate future costs, it is

Figure ES-2. Estimated Cost of Fund-Lead Actions at Current and Future NPL Sites: Base Case, FY 2000–FY 2009 (1999$)

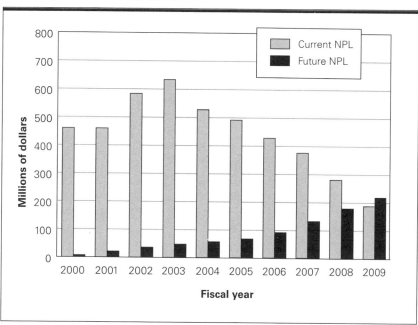

equally important to predict the number of mega sites that will be listed on the NPL and the percentage of actions that will be Fund-lead.

In our interviews, EPA officials anticipated an increase in the number of mega sites likely to be listed on the NPL. The anticipated rise in mega sites is, in part, the result of increased attention to contaminated sediment and mining sites, as well as more focused efforts in a number of states and EPA regional offices to identify the sources contaminating municipal water supplies. In addition, EPA and state managers noted that some costly sites will almost certainly emerge unexpectedly. The average cleanup cost of a mega site is approximately $140 million, more than 10 times that of a nonmega site, which has an average cleanup cost of about $12 million.

Both EPA and state superfund managers expect an increase in the percentage of nonmega sites where cleanup will be paid for with Trust Fund monies. There are two major reasons why this is the case. First, state agencies prefer to address sites under their own programs when responsible parties are financially viable and cooperative—that is, when site cleanups do not require public funds. Second, a typical nonmega NPL site is too expensive for even a well-financed state superfund program to pay for with its own resources. The result is that sites where Trust Fund monies are needed for cleanup are more likely to end up on the NPL.

The availability of state funds to clean up contaminated sites is only one factor that affects whether a state wants a site listed on the NPL. States with less well-financed superfund programs, or no state superfund program at all, often have not been able to muster the political support to place sites on the NPL. On the other hand, states with mature programs and considerable legal and technical expertise often turn to EPA when sites require federal enforcement authority or involve the relocation of residents. Counterintuitively, states generally perceived to have strong, well-funded superfund programs, such as California, Pennsylvania, New Jersey, and New York, have listed the most sites on the NPL.

Forecasting future NPL sites is also complicated by EPA's policy, in place since 1996, of obtaining the governors' concurrence before listing sites on the NPL. Whether a governor wants a site listed depends on various political and economic factors. Thus, even if it were possible to identify all "NPL-caliber" sites not currently on the NPL, that alone is not sufficient to confidently predict the number of future listings.

The cost of the federal Superfund program to state environmental agencies is an increasingly important concern to states. States are liable for two major expenses related to Fund-lead cleanups: 10% of the cost of remedial actions, including long-term response actions (activities aimed at groundwater and surface water restoration), and 100% of the cost of operation and maintenance activities for Fund-lead remedies. States are beginning to feel the financial pinch, especially at mega sites where long-term response actions can cost millions of dollars annually, as more NPL sites reach the latter stages of the cleanup process.

5. Additional information is required to assess the level of resources needed for program management, policy, and administrative support functions to implement the Superfund program.

We found it difficult to understand, much less evaluate, the level of Superfund resources going to program management, policy, and administrative support functions. EPA spends a large portion of its annual Superfund budget on these activities. The costs incurred are primarily associated with personnel and fall into two categories:

- Superfund staff in four EPA headquarters offices: the Office of Solid Waste and Emergency Response, the Office of Enforcement and Compliance Assurance, the Office of Administration and Resources Management, and the Office of the Chief Financial Officer. Within the scope of this study, we were not able to determine the appropriate size of these offices for the future Superfund program.
- Staff in EPA regional offices who work on the Superfund program but do not charge their time to site-specific accounts in the Agency's financial management system. A surprisingly large proportion of regional Superfund staff time is not charged to site-specific accounts, for reasons that are not readily apparent.

Absent better information about what activities are being conducted by these staff, it is difficult to assess where current resources are going and whether current funding levels are appropriate.

6. Post-construction activities are increasingly important to ensure the success of the Superfund program.

Post-construction activities are a crucial part of the Superfund program. These activities include reviews of site remedies—to ensure that they remain protective of human health and the environment—and, for many sites, long-term response action, among other efforts. EPA has historically placed less emphasis on these activities than on remedial cleanup actions. The annual cost to EPA of implementing post-construction activities is relatively small, with the notable exception of long-term response actions at mega sites.

Five-year reviews of site remedies, many of which are required by statute, provide useful information about whether remedies are actually protective and whether additional funds—from either EPA or responsible parties—may be needed in the future. For this reason, we evaluated 151 recently completed five-year review reports.

Our evaluation indicates that EPA needs to improve the quality of these reviews as well as determine whether the recommendations made in the reports have actually been implemented. Ignoring its own guidance, the Agency sometimes concludes that remedies are protective when, in fact, they are incomplete because remedies were not fully implemented, are not functioning as designed, or are unlikely to meet cleanup objectives on schedule. This was the case in almost half of the five-year reviews we examined in which EPA had deemed the remedy at the site protective of human health and the environment.

The results of the five-year reviews provide a useful indication of the future investment that may be needed to ensure that remedies remain protective over time. It is, however, too soon to estimate the magnitude of the future liabilities that will be incurred when some remedies are found to be in need of repair or insufficient.

Recommendations

There are four major issues that we believe need to be addressed by EPA, Congress, and state environmental agencies, in consultation with the full array of Superfund stakeholders. Our recommendations have a common goal: to help formulate a clear mission for the Superfund program and to improve its effectiveness and efficiency.

1. Review and clarify the purpose of the National Priorities List.

It is clear from our interviews with both EPA and state officials that even as the Superfund program enters its third decade, what types of sites should be on the NPL is an unresolved question. That is, the *purpose* of the NPL is unclear. This is not to suggest that the NPL should be so rigidly defined that it precludes regional and state flexibility in determining which sites to list and which sites to address under other cleanup programs. Nevertheless, we recommend that EPA, with the involvement and input of other Superfund stakeholders, should take the following actions.

- Decide whether contaminated sediment and mining sites should be listed on the NPL in greater numbers or whether a separate program or approach is needed. This is an issue that ultimately Congress will have to address, given the potentially dramatic cost implications for the Trust Fund.
- Establish a process for identifying potential NPL sites. Such a process will require coordination with other EPA programs, as well as with state agencies. As part of this effort, EPA should issue periodic reports to Congress outlining recent listing trends, the types of sites the Agency expects to list on the NPL in coming years, and their cost implications.
- Review the governor's concurrence policy to determine whether allowing a state to, effectively, veto the listing of a site that EPA believes should be on the NPL is compromising the effectiveness of the Superfund program.
- Sponsor an independent study of states' financial capacity to pay for state-funded cleanups and their share of Fund-lead actions, as well as the implications that this has on the Superfund program.

2. Assess the level of program management, policy, and administrative support resources needed to implement the Superfund program.

EPA spends a large portion of its annual Superfund budget on management, policy, and administrative support activities. Clearly, these offices and the functions they perform are important. However, to ensure that all resources currently devoted to these efforts are, in fact, needed, we recommend that EPA conduct two studies:

- Evaluate the staffing level and cost of program management, policy, and administrative support functions of the Superfund program to better understand the activities conducted and the staffing and resources required.
- Determine why a large percentage of the activities conducted by staff in the 10 EPA regional offices is charged to nonsite accounts, even though, according to EPA managers, most of the staff spend most of their time working on site-specific activities.

3. Improve the management and financial systems for tracking Superfund progress and costs.

To conduct this study, we relied heavily on two major EPA internal information systems, the Comprehensive Environmental Response, Compensation, and Liability Information System (CERCLIS) and the Integrated Financial Management System (IFMS). We encountered many instances of incorrect data, inconsistent coding, and poor quality control in both systems. In IFMS, the proliferation of codes makes it difficult to use the data to assess the full cost of individual Superfund program actions and activities. We recommend that EPA take the following actions.

- Conduct a rigorous top-to-bottom review of the purpose, structure, and management of CERCLIS and IFMS and evaluate whether these systems are meeting the needs of senior Agency decisionmakers. This review should not be carried out by those responsible for day-to-day management of the systems and should involve outside experts in financial management, information systems, and the Superfund program.
- Revamp CERCLIS to minimize the incidence of incorrect and outdated information.
- Streamline and improve IFMS (and its successor system, now in development) to enhance its usefulness for analyses of the cost of individual components of the Superfund program.

- Improve quality control procedures, communication, and training of users of CERCLIS and IFMS throughout the Agency to ensure greater uniformity in coding practices.

4. Give higher priority to post-construction activities.

EPA, the states, and responsible parties have made significant progress in cleaning up NPL sites. EPA must now ensure that the remedies in place at these sites remain protective of human health and the environment over the long term. To fulfill this obligation, we recommend that EPA take the following steps.

- Review and clarify the definition of the "statement of protectiveness" used in five-year reviews and ensure that the Agency follows its own guidance when evaluating the protectiveness of remedies.
- Create and maintain a system that compiles the recommendations made in five-year review reports and establish a mechanism to make certain that the recommendations are, in fact, implemented.
- Develop a system to track the monitoring and implementation of institutional controls at NPL sites.
- Improve public access to five-year reviews by increasing their clarity and posting the five-year review reports on the EPA website.

* * * * *

We believe that the above recommendations will contribute to a more effective Superfund program in the years to come.

SUPERFUND'S FUTURE

Chapter One

Introduction

The debate about whether and how to reauthorize the Comprehensive Environmental Response, Compensation, and Liability Act of 1980 (CERCLA)—better known as Superfund—has continued relatively unabated since shortly after the law's last major reauthorization, in 1986.[1] Although the underlying debate concerns CERCLA's liability provisions and cleanup standards, over the past few years disagreement has focused on the question: When will the Superfund program start to wind down? More specifically, when will a decrease in activity at National Priorities List (NPL) sites—the sites where the U.S. Environmental Protection Agency (EPA) can fund long-term cleanups from the Hazardous Substance Superfund (Trust Fund)—translate into decreased annual appropriations for EPA as part of the Departments of Veterans Affairs and Housing and Urban Development, and Independent Agencies (VA-HUD) appropriations bill?

The emergence of this question suggests a major turnaround for Superfund. Historically, EPA has faced persistent criticism for its implementation of the Superfund program, particularly for the pace of cleanup and the selection of remedies. Now, a number of EPA's harshest critics are singing the Agency's praises. Diverse organizations, such as the U.S. General Accounting Office (GAO) and the National Academy of Public Administration, as well as a group of private attorneys brought together under the auspices of the Information Network for Superfund Settlements, have recently lauded the success of the program, specifically the administrative reforms started under former EPA Administrator William K. Reilly and continued under his successor, Carol Browner.[2]

Much of the praise for EPA reflects the progress that has been made toward cleaning up sites on the NPL. By the end of FY 1999 just over half (52%) of the sites on the NPL (not including federal facilities) had been designated "construction-complete" by EPA, which means that the "physical construction

1

of all cleanup actions are complete, all immediate threats have been addressed, and all long-term threats are under control."[3] By the end of FY 2000, the number of construction-complete sites on the NPL (not including federal facilities) had increased to 728, or 57%.

Notwithstanding this progress, much work remains to be done. For instance, at the 553 NPL sites that were not construction-complete at the end of FY 2000, hundreds of cleanup remedies had yet to be started. In addition, many of the 728 sites that were construction-complete at the end of FY 2000 still had unfinished remedies. Moreover, even sites that have all their remedies in place may require years, if not decades, of operation and maintenance activities to attain cleanup goals or to ensure the long-term protectiveness of public health and the environment. In addition to the remaining work at existing NPL sites, of course, is the cleanup to be conducted at sites added to the NPL each year.

Whether the Superfund program will soon "ramp down" is a pressing concern to many stakeholders. Currently, a wary truce exists between those who would leave CERCLA's liability and cleanup provisions intact and those who do not want to reinstate the authority for the taxes that filled the coffers of the Trust Fund, which expired at the end of 1995.* Meanwhile, there are other interests who are seeking a major overhaul of the law's liability and cleanup provisions.

These conflicting visions of Superfund's future may soon come to a head. At the end of FY 2000, the Trust Fund had an unappropriated balance of approximately $1.3 billion,[4] the equivalent of less than one year of the program's current spending. Although nothing precludes Congress from continuing to authorize funds for the Superfund program, even if the balance in the Trust Fund is zero, questions about the long-term funding needs of the program will inevitably take center stage and be part of any debate about the reauthorization of the statute.

Request for RFF Report

In spring and summer 1999, congressional staff on the Senate and House Appropriations Committees, as well as on the Senate Environment and Public Works and House Commerce and Transportation and Infrastructure Committees, and senior EPA officials, tried unsuccessfully to determine how much money the Superfund program would need over the next few years and when

---

*The Trust Fund is stocked primarily by three taxes—a petroleum tax, a chemical feedstocks tax, and a corporate environmental income tax—as well as by general revenues.

a decrease in needed appropriations was likely to occur. Participants agreed that a more detailed analysis was required to answer these questions. As a result, the conference report to the FY 2000 VA-HUD appropriations bill included a charge to EPA to commission Resources for the Future (RFF) to conduct an independent study to help answer the seemingly simple question: How much money is needed to fund the Superfund program each year from FY 2000 through FY 2009? The conference report reads,

> Finally, the conferees direct that within 45 days of enactment of this Act, EPA award a cooperative agreement for an independent analysis of the projected federal costs over the ten year period of fiscal years 2000-2010 [sic] for implementation of the Superfund program under current law, including the annual and cumulative costs associated with administering CERCLA activities at National Priorities List (NPL) sites. The analysis should identify sources of uncertainty in the estimates, and shall model: (1) costs of completion of all sites currently listed on the NPL, (2) costs associated with additions to the NPL anticipated for fiscal year 2000 through fiscal year 2009, (3) costs associated with federal expenditures for the operations and maintenance at both existing and new NPL sites, (4) costs for emergency removals, (5) non-site specific costs assigned to other activities such as research, administration, and interagency transfers, and (6) costs associated with five-year reviews at existing and new NPL sites and associated activities. For purposes of this analysis, costs associated with assessment, response, and development of brownfields and federal facility sites are not to be included. The analysis shall be conducted by the [sic] Resources for the Future...[5]

The congressional language specifically excludes from the RFF study estimates of the future cost of cleaning up sites owned and operated by federal agencies, such as the Department of Energy and the Department of Defense, referred to as federal facilities, as well as the cost incurred by EPA for overseeing federal facility cleanups. Estimates of EPA's cost for the brownfields program are also excluded. Accordingly, in the discussion that follows, we exclude all EPA expenditures associated with federal facilities and the brownfields program.*

---

*In FY 1999, EPA spent about $32 million on activities related to federal facilities and about $24 million on the brownfields program. This low level of expenditures in the brownfields program is the result of including only expenditures, rather than obligations. In FY 1999, the Agency obligated more than $90 million to the brownfields program, but much of this money was not to be spent until after FY 1999, as it is in the form of multiyear grants and cooperative agreements.

Although the conference report identifies six elements, estimates of the future cost to EPA of the Superfund program hinge, fundamentally, on three key questions.

1. How much money will be needed to complete cleanup activities at those sites currently on the NPL?
2. How many sites will be added to the NPL over the next 10 years, and what are the costs to EPA associated with addressing these sites?
3. What costs will EPA incur from FY 2000 through FY 2009 for other Superfund activities paid for with Trust Fund dollars, including the removal program; site assessment; program staff, management, and support; program administration; and other programs and agencies that undertake Superfund-related work?

Several aspects of the task assigned to RFF are important to note. First, our estimates are to represent the cost of implementing the program "under current law"—in other words, assuming no change in the law's liability or cleanup standards and no major change in current EPA policies. Another implicit assumption is that our estimates of future costs are not constrained by past funding levels. That is, our goal is to model the work of the Superfund program and, based on that, derive estimates of the funds needed. Thus, we are not limited in our estimates by historical levels of congressional appropriations, but neither do we perform an all-out "needs assessment" of the Superfund program, which would require an appraisal, on a national scale, of the number of sites and the extent of contamination across the country, as well as an evaluation of how much cleanup is needed.

The congressional language also specifies that only the cost of the Superfund program to EPA are to be included. Thus we do not address the cost of the Superfund program incurred by potentially responsible parties (PRPs)* or by the states, which have an important role in the program's implementation.[6] Although the states are responsible for 10% of the total cost of remedies paid for with Trust Fund monies and 100% of the cost of the operation and maintenance of these remedies, we do not estimate how the pace of NPL cleanups will affect the need for state funds. Clearly, this is an important issue not only to the states but also to the overall success of the Superfund program. We do,

---

*Potentially responsible parties are those who are liable under Superfund for site cleanup. They typically include private companies, state or local governments, and in the case of federal facilities (as well as other sites), agencies of the federal government.

however, include in our estimates the costs incurred by other federal agencies (for example, the cost of Superfund enforcement borne by the Department of Justice) when these dollars are appropriated through EPA's Superfund budget.

Another result of the congressional language is that we focus on EPA *expenditures*. Most descriptions of the cost of the Superfund program have used EPA budget numbers, or what are referred to as obligations. Budget and obligations figures are basically the amounts allocated to different elements of the Superfund program at the beginning of the fiscal year. They reflect EPA priorities for different functions, as well as what EPA thinks different program elements are likely to cost. The amount of money actually spent each year on these functions, however, can differ significantly from beginning-of-the-year expectations. Since our goal is to estimate the likely costs of various functions in future years, throughout this report we use *only* expenditure data. Thus, our estimates for the annual needs of the Superfund program reflect the amount of money EPA will actually spend. These expenditure data will need to be translated into obligations figures for the appropriations process.

Because our focus is on the actual annual cost to EPA and the Trust Fund, we also do not include information on cost recovery dollars—that is, funds that are recovered from responsible parties by EPA and the Department of Justice. Cost recovery dollars are returned to the Trust Fund but, in general, must be reappropriated to EPA by Congress and thus do not reduce the need for annual appropriations to the Agency.[7]

Finally, the congressional report language directed RFF to address uncertainty in its cost estimates. We do this by developing three scenarios: a base case, a low case, and a high case. We believe that the three scenarios are realistic. In other words, we have not, as is sometimes done, artificially inflated the high estimate and deflated the low estimate to make our middle estimate, or base case, look good. That said, however, our judgement, based on the full body of research, analyses, and interviews conducted for this study, is that the base case is the best estimate of future costs.

Overall Approach and Data Sources

Our overall approach to estimating the future cost of the Superfund program, as dictated by the congressional report language, is to build a model of future Superfund costs. By "model" we mean that we identify the major elements of the program, assign a specific cost to each, scale associated program costs accordingly, and predict when these expenditures will be incurred in each of 10 fiscal years, from FY 2000 through FY 2009.* Throughout the report we try to make our assumptions transparent. Although FY 2000 has come and gone while this report was being written, we present our estimates for FY 2000, as expenditure data for FY 2000 were not available in time to be incorporated.

We have chosen to keep our model relatively simple for three reasons. First, because the data sources that we relied on were not very sophisticated, it seemed inappropriate to create a fancy model with lots of bells and whistles. Second, we wanted to be able to explain our assumptions and results in plain English. Finally, our goal was to create a model (and cost estimates) grounded in the real world, one that closely reflects the way the Superfund program is actually implemented under the current statute. Thus, we do not attempt to model the cost of an "ideal Superfund program"; rather, our model takes into account the many vagaries of the program.

Our most intensive efforts went into estimating the future cost of addressing NPL sites by (1) developing average costs for key phases of the NPL cleanup process paid for with Trust Fund monies and (2) developing alternative scenarios for estimating the number, type, and cost of sites likely to be added to the NPL in the coming years. We also spent considerable time collecting information about the removal program, which is the other core cleanup program implemented under Superfund. We dedicated less time to understanding the intricacies of the day-to-day costs of running the Superfund program (for example, the costs associated with policy staff at EPA headquarters and management in the 10 EPA regional offices). These costs represent a significant portion of the total costs of the Superfund program and deserve greater scrutiny than time and resource constraints allowed.

At the beginning of the study we received large datasets from two EPA databases: CERCLIS, the Comprehensive Environmental Response, Compensation, and Liability Information System, which is EPA's major Superfund tracking database; and IFMS, the Integrated Financial Management System, which is EPA's internal accounting system that tracks all individual expenditures. Both datasets were complete through the end of FY 1999.

We used CERCLIS information to create a baseline of the current status of all operable units (individual projects at individual sites) at all nonfederal NPL sites. At the 1,245 final and deleted sites on the NPL at the end of FY 1999, there were approximately 2,200 operable units. This was our starting point for

---

*The model is described in detail in Appendix G.

determining, as of the beginning of FY 2000, the number of cleanup actions that had been completed, the number that were under way, and the number not yet started at all nonfederal NPL sites.

IFMS is the source of all our historical expenditure data and provides information on the actual costs of different parts of the Superfund program—information that is more reliable for making estimates than the budget numbers developed at the beginning of each fiscal year. However, it quickly became clear that there are inconsistencies in how data are coded, how EPA staff bill (or charge) their time, and how certain expenses are categorized.* These inconsistencies make the data difficult to use for certain types of analyses. GAO has raised these issues in numerous reports.[8] The first problem we encountered was simply understanding the Agency's system for coding expenditures. It became apparent that certain costs—such as expenditures for remedial actions—are coded relatively consistently; other costs—such as expenses for five-year reviews—are not.

We have no reason, based on our work, to question the validity of the costs charged to specific sites. We are concerned, however, that the coding inconsistencies result in the underreporting of some site-specific costs. We spent months reviewing EPA data and met with EPA staff and officials to determine what data were appropriate for our task and where we needed to gather new information. We did not—as this would be a major study of its own—evaluate the underlying basis for the financial data provided by EPA.†

We used IFMS data in two ways. First, we analyzed data on past expenditures to develop average costs—what we refer to as unit costs—for major cleanup actions at NPL sites. For most of our estimates of unit costs, we used data in IFMS from FY 1992 through FY 1999. We focused on this period because there is general agreement that by FY 1992 the Superfund program had matured and because this time span provided us with sufficient data for analysis. Second, we used expenditure data in IFMS for FY 1999 as our baseline

---

*Our analyses of IFMS data did not include an evaluation of EPA's indirect cost accounting methodology, and nothing in this report should be taken as either support for or criticism of that methodology or any of its components.

†Our critique of IFMS data is based solely on the purposes for which we used the data and not for other purposes for which IFMS data are used, such as cost recovery. According to EPA and the Department of Justice, all site-level data in IFMS are supported by written documentation, but review of the documentation was impracticable as part of this effort.

for estimating the costs of many of the other major elements of the Superfund program, such as staff time, management, and rent.

In our model, we grouped individual NPL sites into several categories. The most important distinction, for our purposes, is between sites with cleanup costs of $50 million or more, referred to as mega sites, and sites with direct cleanup costs of less than $50 million, referred to as nonmega sites.[9] We developed the list of mega sites by asking the 10 EPA regional offices to identify those sites on the NPL that had—or were expected to have—total removal and remedial action costs of $50 million or more. The mega site determination is applied both to sites cleaned up directly by EPA, referred to as Fund-lead sites, and to sites cleaned up by responsible parties, referred to as PRP-lead sites. Of the 1,245 sites on the NPL at the end of FY 1999, EPA designated 112 as mega sites.*

Because of the importance of mega sites in determining the future cost of the Superfund program, we gathered site-specific information on each of these sites. We asked the remedial project managers (RPMs) in the relevant EPA regional offices for detailed information on the number, types, and likely timing of all current and future cleanup actions at mega sites, and whether these actions were, or were likely to be, paid for by EPA or by responsible parties.† In addition, we asked the RPMs to estimate the cost of all actions likely to be paid for by EPA. To inform our estimates of the number and type of sites likely to be added to the NPL in the future, we also conducted telephone interviews with the Superfund division directors of all 10 EPA regional offices‡ and with hazardous waste officials from nine states.[10] Our purpose was to gather specific information on the sites likely to be added to the NPL in the coming years, as well as to gather information on the capabilities of state programs.**

In addition to these data collection efforts, we conducted many interviews with EPA staff in headquarters, as well as with senior officials at the Department of Justice and elsewhere, to gather information about the cost of the Superfund program. We also submitted innumerable questions to EPA, seeking cost information, historical data, and clarifications of Agency policy and guidance.

---

*See Appendix D for a list of these sites.

†EPA considers some of the data we collected "enforcement confidential," since it indicates whether the Agency believes it will be able to obtain a settlement with the responsible parties. As a result, RFF is not making public this information except in aggregate form.

‡See Appendix A for the list of the 10 EPA regional offices and the states they include.

**See Appendix E for details on our interviews regarding future NPL sites.

Reliability of the Estimates

It is important, at the outset, to acknowledge the uncertainty of the estimates presented in this report—an issue anticipated in the conference report language. First and foremost, we are limited by the quality of the data available and of course by our ability to predict the future. We had no choice but to rely on existing EPA information, although as already noted, we did collect new information in a number of key areas.

Even if we had perfect data, however, we cannot stress too much that *these are estimates, and only estimates.* It is highly unlikely that we have predicted the exact cost of the Superfund program in any specific year, much less 10 years from now. In a commentary on National Public Radio, former Secretary of Labor Robert Reich discussed 10-year budgets and stated the problem succinctly: "The fact is ten-year budget projections are totally meaningless. There are just too many things that could happen over ten years, even tiny things, that will knock a ten-year budget projection completely out of whack."[11]

Nevertheless, the results of our research should be useful to policymakers. First, some components of our estimates are likely quite reliable. For example, the actual costs of completing cleanup at sites currently on the NPL will likely fall within the range of our estimates. These costs constitute the lion's share of direct cleanup costs through FY 2009. Second, studies like this one can identify which factors are most important in estimating future costs. This informs appropriators on Capitol Hill and other policymakers which parts of the Superfund program to watch as predictors of future costs, and it suggests the cost implications of different policy choices. Third, most estimates of future costs are likely to be more accurate in the first few years than in the later years. This applies to our estimates as well. Finally, we believe that the basic shape of the "cost curve" we have developed accurately represents the cost of the Superfund program from FY 2000 through FY 2009.

Six Program Elements

Over the past 10 years Congress has appropriated roughly between $1.3 billion and $1.6 billion each year to EPA for the Superfund program.[12] As shown in Figure 1-1, EPA Superfund expenditures have closely mirrored these annual appropriations.

When we began this study, we had hoped to be able to assign all costs to specific program functions. For example, we would have liked to attribute to

Figure 1-1. Total Annual EPA Superfund Expenditures, FY 1990–FY 1999
(Nominal Dollars)

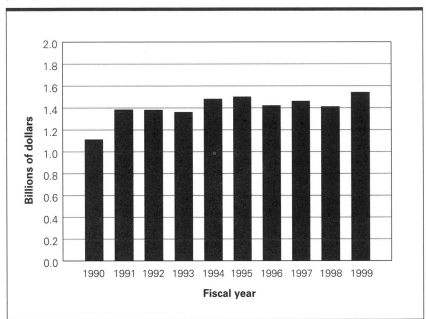

Note: Expenditures associated with federal facilities and brownfields are not included.
Source: RFF IFMS dataset.

the removal and remedial programs "their share" of rent, financial management activities, and regional staff costs not charged directly to sites. We were unable to do this because of the way costs are coded in IFMS. An exhaustive audit would have been needed to determine the true fully loaded costs of each major Superfund function. Such an audit was beyond the scope of this study.

Instead, we developed six program elements to group Superfund program costs:

- *removal program:* cleanup program responsible for emergency responses and short-term cleanup actions at NPL and non-NPL sites;
- *remedial program:* cleanup program responsible for long-term cleanup actions at NPL and non-NPL sites;
- *site assessment:* activities undertaken to evaluate the severity of contamination at sites and to determine what, if any, response should be conducted;
- *program staff, management, and support:* staff payroll and benefits for general management and support activities associated primarily with the

remedial and enforcement programs, response action management, state programs and grants, laboratory analysis, and various EPA offices;

*   *program administration:* rent, facilities operation and maintenance, equipment and supplies, activities carried out by the EPA Office of Administration and Resources Management, regional administrative and resource management staff time, and activities performed within the EPA Office of the Chief Financial Officer; and

*   *other programs and agencies:* Superfund-related work of offices within EPA, including the Technology Innovation Office, the Chemical Emergency Preparedness and Prevention Office, and the Office of the Inspector General, and other federal agencies, including the National Institute for Environmental Health Sciences and the U.S. Coast Guard.

These six program elements do not represent discrete categories, as there is no simple way to disentangle the interconnected parts that make up the Superfund program. As a result, we were not always able to treat costs within the categories in exactly the same way. However, the level of detail we present in the body of the report and in the appendixes should provide sufficient information for those interested in regrouping these costs. The FY 1999 costs of each of these six program elements are shown in Figure 1-2. We refer to these program elements throughout the report and present our estimates for each from FY 2000 through FY 2009 in Chapter 7.

Outline of the Report

In this report, we discuss each of the six elements of the Superfund program identified by Congress. The bulk of our effort went toward understanding and estimating the direct cleanup costs that are encompassed in the two major EPA cleanup programs—the removal program, discussed in Chapter 2, and the remedial program, discussed in Chapters 3 and 4. These two programs form the core of the Superfund program. Chapter 3 includes a description of the cleanup process that typifies EPA's approach to addressing NPL sites and also discusses some of the key building blocks for estimating future costs. In Chapter 4 we describe post-construction activities, which are an increasingly important component of the Superfund program. Post-construction activities are the efforts required at NPL sites after the physical or engineering components of a remedy have been implemented, many of which are critical to protecting public health and the environment.

Figure 1-2. FY 1999 Expenditures for the Superfund Program

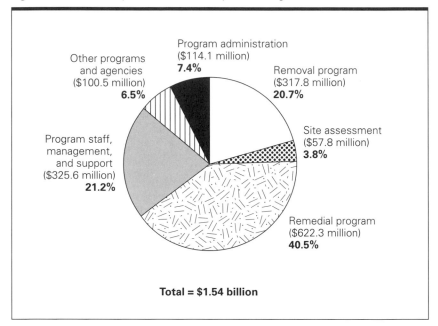

Program administration
($114.1 million)
7.4%

Other programs
and agencies
($100.5 million)
6.5%

Removal program
($317.8 million)
20.7%

Program staff,
management,
and support
($325.6 million)
21.2%

Site assessment
($57.8 million)
3.8%

Remedial program
($622.3 million)
40.5%

Total = $1.54 billion

Note: Expenditures associated with federal facilities and brownfields are not included.
Source: RFF IFMS dataset.

In Chapter 5 we address the question that is certainly the most controversial aspect of this study: How many sites are likely to be added to the NPL in the future? In this chapter we describe the program to list sites, analyze historical data on listings, and summarize what we learned from interviewing staff at the 10 EPA regional offices and at 9 state environmental agencies about the factors that lead both EPA and the states to pursue listing sites on the NPL. We conclude Chapter 5 with a description of three alternative scenarios used in our model to estimate the possible range of costs of sites added to the NPL from FY 2000 through FY 2009.

In Chapter 6 we describe the costs incurred by EPA for Superfund program support activities. Included in this category are activities specifically associated with the removal and remedial programs, such as regional staff payroll and benefits, and general program management, support, and administration. This chapter also includes a brief description of other programs in EPA and other federal agencies whose Superfund-related work is funded with monies from the Trust Fund.

In Chapter 7 we present our model and our estimates of the future cost of the Superfund program, layer by layer. We begin with our estimates for the future cost of the removal program. We then provide our estimates for the remedial program, including the cost of Fund-lead cleanup actions at current and future NPL sites, other site-specific activities at NPL sites (oversight, enforcement, response support, community and state involvement, and five-year reviews), and efforts undertaken at non-NPL sites. Next, we project the cost of program staff, management, and support; program administration; and other programs and agencies financed by the Superfund program. And finally, we sum the estimates for each component to arrive at the total cost to EPA over the next decade.

Appendixes provide details on our methodology and supplemental information.

Chapter Two

The Removal Program

The removal program is the part of the Superfund program responsible for conducting emergency responses and short-term cleanup actions. Removal actions can be undertaken at sites irrespective of whether they are on the National Priorities List (NPL) and are most often taken to address the release or threat of release of hazardous substances in time-sensitive situations. The removal program accounted for nearly 21% of total Superfund expenditures in FY 1999 and serves a different purpose from that of the remedial program.

In this chapter, we provide an overview of the removal program, beginning with a description of the U.S. Environmental Protection Agency's (EPA) statutory authority to conduct removal actions, the types of removal actions, and the historical use of removal actions. We then discuss what is currently known about the cost of the removal program, paying special attention to the "extramural" dollars spent by EPA for removal actions, that is, expenditures made by EPA that are external to the Agency. In the final section of this chapter, we discuss our approach to estimating the cost of the removal program from FY 2000 through FY 2009.

Overview of the Removal Program

The removal program grew out of an early recognition by EPA that some situations involving the release or threat of release of hazardous substances would require a quick response. The original authority for the program came from the 1972 Clean Water Act, which gave EPA the power to respond to oil and hazardous substance releases. This authority was subsequently expanded in 1980 with the enactment of the Superfund statute, which gives EPA the authority to conduct removal actions whenever there is a release or threatened release of a hazardous substance into the environment, or a release or threatened release of a pollutant or contaminant that may present an imminent and substantial danger to the public health or welfare.[1]

Removal actions are defined in section 101(23) of the Comprehensive Environmental Response, Compensation, and Liability Act of 1980 (CERCLA) and in section 300.5 of the National Oil and Hazardous Substances Pollution Contingency Plan (referred to as the National Contingency Plan). As summarized in EPA guidance, removal actions include the following:

- actions taken in the event of a release or threatened release of hazardous substances into the environment;
- actions taken to monitor, assess, and evaluate the release or threatened release of hazardous substances;
- disposal of removed material; and
- other actions to prevent, minimize, or mitigate damage to the public health or welfare or to the environment.[2]

CERCLA limits removal actions carried out by EPA and paid for with Trust Fund monies to 12 months and $2 million,[3] although these limits can be exceeded if a waiver is obtained. In fact, according to EPA analysis, since the beginning of the Superfund program, about 8% of the removal actions conducted by the Agency have cost more than $2 million.[4] The time and monetary limits do not apply to removal actions carried out directly by responsible parties.

The removal program serves a need not fulfilled by the remedial program. Whereas the remedial program focuses mostly on NPL sites, the majority of removal actions are conducted at sites that are not, and may never be, listed on the NPL. From FY 1992 through FY 1999, approximately 76% of removal actions were performed at non-NPL sites,[5] and an additional 2% were undertaken at proposed NPL sites, that is, sites not yet final on the NPL.

EPA staff in the removal program can move quickly to address a release or threatened release of hazardous materials, stabilizing the site and in some cases minimizing the need for future cleanup action. As shown in Table 2-1, at the vast majority (87%) of sites nationwide at which the release or threatened release of hazardous materials has been addressed, only one removal action has been required. Six to 10 removal actions have been conducted at a handful of sites, and at one large mining site in Colorado, California Gulch, more than 60 removal actions have been undertaken since the beginning of FY 1992.

Although the focus of this chapter is the role and cost of removal actions in the context of the Superfund program, EPA's responsibilities to carry out emergency responses exist under other statutory requirements as well, as sum-

Table 2-1. Removal Actions per Site, FY 1992–FY 1999

Number of removal actions per site	Number of sites
1	1,781
2	196
3	44
4	16
5	7
6	2
7	2
8	0
9	1
10 or more	4
Total	**2,053**

Source: Removal action starts data provided to RFF by EPA, March 2000.

marized in Table 2-2. Many of these programs are not funded with Superfund monies, such as Oil Pollution Act activities and counter-terrorism efforts,[6] but the growth of such responsibilities does increase the overall workload for removal program staff.

Types of Removal Actions under Superfund

For purposes of setting priorities, defining the nature and degree of response, and allocating resources, EPA has defined three types of removal actions:

- *emergency responses:* removal actions where the release or threat of release requires that response activities begin on-site within hours of the lead agency's determination that a removal action is appropriate;
- *time-critical actions:* removal actions where, based on a site evaluation, the lead agency determines that a removal action is appropriate and that on-site activities must be initiated within six months; and
- *non-time-critical actions:* removal actions where, based on a site evaluation, the lead agency determines that a removal action is appropriate and that a planning period of at least six months is available before on-site activities

Table 2-2. Statutes and Mandates Requiring the Involvement of EPA's Removal Program

Statute/mandate	Date enacted	EPA responsibilities
CERCLA	1980	• Emergency response • Time-critical and non-time-critical removal actions • Development and implementation of regional contingency plans • Establishment and maintenance of system for receiving and evaluating all notifications of hazardous substance releases
Oil Pollution Act	1990	• Response to oil spills
Robert T. Stafford Disaster Relief and Emergency Assistance Act (Stafford Act)	1988	• Emergency response to hazardous materials releases resulting from natural disasters (hurricanes, tornadoes, floods, earthquakes, etc.)
Presidential Decision Directives on Counter-terrorism	1995, 1998	• Protection of EPA personnel and facilities against terrorism • Support to Department of Justice, Federal Bureau of Investigation, and Department of Defense (training, equipment, specialized units) • Support to the Federal Emergency Management Agency (hazmat and environmental consequence management) • Protection of water supplies • Vulnerability awareness and educational program

must begin. The lead agency for non-time-critical removal actions will undertake an engineering evaluation/cost analysis or its equivalent.[7]

Time-critical actions are unquestionably the most common type of removal action conducted by EPA. In FY 1999, for example, the Agency conducted 314 Fund-lead removal actions, of which 210 (67%) were time-critical actions, 77 (25%) were emergency responses, and only 27 (9%) were non-time-critical actions.*

The urgency typically associated with emergency responses and time-critical removal actions means that the approach to conducting these actions differs substantially from that required for non-time-critical actions. Regional staff known as on-scene coordinators (OSCs) have primary decisionmaking

---

*Percentages do not add because of rounding. EPA conducted four additional removal actions, but data on their type were missing from the "Removals historical summary data" provided to RFF by EPA, March 2000. (See Appendix C for details.)

authority for most emergency responses and time-critical actions when EPA is the lead agency.[8] Non-time-critical removal actions, in contrast, must follow procedures that more closely parallel the remedial process, with a required site investigation study (a streamlined version of the remedial investigation/feasibility study required at NPL sites) and public comment period. Remedial project managers, the OSCs' counterparts in the remedial program, are increasingly conducting non-time-critical removal actions as part of the remedial program.

Table 2-3 highlights examples of the different types of actions EPA takes under the removal authority granted by CERCLA. These examples come from Agency publications and our interviews with the regional program managers. Since EPA's response is dictated largely by the urgency—the release or threat of a release—of the particular situation, similar activities may be conducted as part of different removal types.

A Historical Perspective

EPA's use of its authority under CERCLA to conduct removal actions has grown substantially in the two decades since Superfund was enacted. There was a 70% increase in the number of removal actions conducted by the Agency in the 1990s compared with the decade before.[9] As shown in Figure 2-1, from FY 1992 through FY 1999, the total number of removal actions conducted by EPA has remained relatively constant, averaging about 315 a year.[10] The proportion of emergency response, time-critical, and non-time-critical removal actions conducted annually has also varied little during this period.

Among the EPA regions, considerable variation exists in the number and type of removal actions conducted. From FY 1992 through FY 1999, Regions 2, 4, and 5, for instance, conducted 367, 350, and 412 removal actions, respectively, while Regions 6, 9, and 10 conducted only 182, 138, and 79 removal actions, respectively. Figure 2-2 shows the regional breakdown of removal actions by each of the three types.

The regional program managers indicated in our interviews that they chronically have more sites requiring emergency responses or time-critical removal actions than they have the resources to address, although no documentation for this shortfall exists. As new sites come to the attention of the Agency, managers are required to juggle their priorities to address the most critical situations first. Non-time-critical actions are often too complex and costly to be undertaken in the removal program, as a single site might require a large portion of a region's

Table 2-3. Examples of Emergency Responses and Removal Actions Conducted by EPA

Typical actions	Examples
Emergency response	
• Response at truck accidents and train derailments involving chemical releases • Provision of bottled water or water treatment systems for sites with contaminated water supplies • Removal and disposal of drums or chemicals abandoned by roadsides or in vehicles • Response to fires and explosions involving chemicals at operating or abandoned facilities • Cleanup and monitoring of mercury contamination at schools and private residences where children have obtained and played with metallic mercury • Response to tire fires • Response to grain elevator explosions involving pesticide releases • Cleanup and monitoring in response to chemical releases resulting from natural disasters (tornadoes, floods, hurricanes) • Response to hazardous releases from abandoned mines to surface waters and/or drinking supplies	**Shenandoah Road Site, Fishkill, New York:** A family discovered their well to be contaminated with volatile organic chemicals (VOCs) at concentrations 20 times the EPA's maximum contaminant level for VOCs in drinking water. EPA Region 2 personnel conducted sampling that determined the well to be one of many contaminated by an extensive plume from an unknown source. As an emergency response, EPA provided activated carbon systems at 52 homes. The state has requested NPL listing for the site. **Divex Site, Columbia, South Carolina:** On September 6, 1993, an explosion severely damaged the Divex explosives manufacturing facility, killing the owner. EPA emergency response personnel responded and found the manufacturing facility extremely unsafe and threatened by further explosions. The site included many explosive devices and detonators, shock-sensitive chemicals, and explosive residues. EPA, with the Bureau of Alcohol, Tobacco, and Firearms and the Department of Defense, temporarily relocated four families living adjacent to the facility and removed or detonated hazardous and explosive materials.

continued on next page

removal budget. Consequently, most non-time-critical removal actions at NPL sites are funded from the remedial program budget.

In our interviews, 7 of the 10 EPA regional program managers of the removal program characterized the federal program as "a critical safety net" for the states and territories in their jurisdiction. Although states have been expanding their emergency response and time-critical removal capabilities, as was made clear in a recent survey conducted by the Association of State and Territorial Solid Waste Management Officials (ASTSWMO), state capacity still

Table 2-3. *continued*

Typical actions	Examples
Time-critical	
• Restriction of access to and removal and disposal of chemicals at abandoned or bankrupt facilities and warehouses that may be subject to vandalism and fires; examples of such facilities include small electroplating shops, paint manufacturers, waste oil "recycling" facilities, and methyl amphetamine labs • Responses to the threat of releases from secondary lead smelters • Cleanup of methyl parathion contamination from illegal applications of the pesticide at individual residences • Stabilization of mining wastes to prevent releases of toxic materials to surface and ground waters	**Interroyal Site, Plainfield, Connecticut:** A large mill in a primarily residential area had been used for the manufacture of furniture. The company was supposed to conduct cleanup under a state agreement but filed for bankruptcy in 1987. Drums, vats, and other containers of acids, bases, and metal-plating solutions were left in the building. The state requested that EPA take a removal action at the facility in 1995. **Mesh Plastics Site, Lake Charles, Louisiana:** After a plastics manufacturing facility was abandoned in 1993, the state found that Mesh Plastics had been operating an illegal, unpermitted hazardous waste storage facility that held approximately 70 drums of acetone and other chemicals. During the assessment, the region determined that although most of the drums were in fairly good shape, a few showed signs of deterioration. The building's roof was also found to be leaking and there was no working fire suppression system.
Non-time-critical	
• "Hot spot" removal • Installation of groundwater treatment systems for contaminated groundwater in conjunction with the remedial program	**Mohonk Site, New Paltz, New York:** EPA installed a groundwater "pump-and-treat" system with air strippers and carbon polishing to treat groundwater from a contaminated plume that had advanced one mile from its source, an old paint manufacturing facility. Approximately 80 residential wells had been contaminated. The system is now being taken over by the remedial program.

Sources: Interviews with EPA Region 2, July 2000; information provided to RFF by EPA, August 2000.

varies. Even the states with the most advanced programs do not have the capacity to address on their own all the emergency responses and time-critical removal actions that arise each year. This finding was confirmed by most of the regional program managers and is supported by ASTSWMO, which recently recommended the preservation of a strong federal role in providing these services.[11]

Figure 2-1. Removal Actions Conducted by EPA by Type, FY 1992–FY 1999

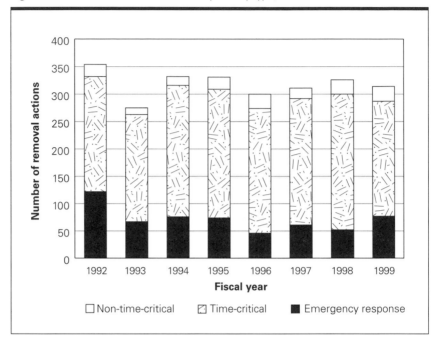

Source: Removal action starts data provided to RFF by EPA, March 2000.

Cost of the Removal Program

EPA spent $317.8 million—$34 million at EPA headquarters and $283.8 million in the EPA regional offices—for the removal program in FY 1999. The costs incurred at EPA headquarters are predominantly for program management and staff. In contrast, the costs incurred by the EPA regional offices are primarily for site-related activities, especially the costs associated with Fund-lead removal actions (mostly contracts with the U.S. Army Corps of Engineers or private contractors). Only a small portion (3%, or $7.6 million) of the dollars spent in the 10 EPA regional offices funded management and support activities.[12]

Approximately $276.2 million—almost 87% of all FY 1999 removal expenditures—went to site-related activities. Of this total, about $220.4 million was incurred for extramural and staff (and other intramural) costs for removal actions carried out or financed by EPA. This $220.4 million is often thought to represent the total annual cost of the removal program, but it is only one (albeit the largest) component.

Figure 2-2. Removal Actions Conducted by EPA by Type and Region, FY 1992–FY 1999

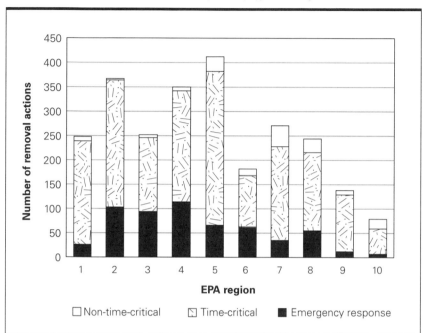

Source: Removal action starts data provided to RFF by EPA, March 2000.

The balance of site-related costs, $55.9 million, went to a number of site-specific activities, including the following:

- response support, which are activities supporting the removal program, such as analytical and laboratory work to detect and measure contamination at sites, contract management, site security and maintenance, health assessments, technical assistance, and other general support and management efforts ($28.1 million);
- enforcement activities carried out by EPA and the Department of Justice, such as identifying potentially responsible parties (PRPs), negotiating with responsible parties to pay for removal activities, developing enforcement actions, and bringing cost-recovery actions ($14.7 million);
- EPA oversight of removal actions conducted directly by responsible parties, which includes direct on-scene oversight of PRP response work, review of reports, work products, health and safety plans, and other PRP documents, and review of laboratory analyses and groundwater monitoring reports ($7.6 million); and

- community and state involvement, which includes activities to facilitate local public involvement and information efforts, management assistance to enable states to participate in a supporting role at EPA-lead cleanups, and state core grants to enhance state site-specific technical cleanup capability ($5.5 million).

These activities are integral to the successful cleanup of sites under both the removal program and the remedial program.

Because extramural spending constitutes such a large proportion of total removal expenditures, it provides a useful measure of historical spending for the removal program. As shown in Figure 2-3, from FY 1992 through FY 1999, there was fluctuation in annual extramural spending for the removal program. The rise in spending during the first part of this period in part reflects the increased use of removal actions after the initiation of the Superfund accelerated cleanup model (SACM) in 1991, which encouraged the use of "early actions" (i.e., removal actions) to quicken the pace of cleanup at NPL sites.

Figure 2-3. Removal Program Extramural Expenditures, FY 1992–FY 1999 (Nominal Dollars)

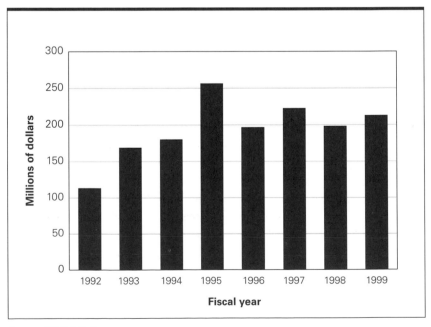

Source: RFF IFMS dataset.

Between FY 1995 and FY 1999, annual extramural spending ranged from about $200 million to $250 million, with a five-year average of approximately $216 million. According to EPA, the peak in FY 1995 likely represents disbursement of funding allocated from a $50 million set-aside from the remedial action budget, which was part of the SACM initiative. A number of sites with expensive removal actions received significant amounts of funding from this set-aside in FY 1995 (e.g., Summitville Mine in Colorado, Ralph Gray Trucking Co. in California, and the Stratford Asbestos site in Connecticut).

The smaller increase in spending in FY 1997 largely reflects emergency response actions taken to address methyl parathion contamination in Jackson County, Mississippi, and in New Orleans, Louisiana. According to EPA data, the Agency spent nearly $40 million to respond to methyl parathion in Jackson County over two and a half years and more than $17 million in New Orleans over three and a half years; a large portion of these expenditures was spent in FY 1997.[13]

Most removal actions are initially, if not ultimately, paid for with Trust Fund monies, which reflects the limited time EPA generally has to involve PRPs. As Table 2-4 indicates, the greater the degree of PRP involvement, the less time-sensitive the action. Virtually all recent emergency responses have been Fund-lead.

Predicting the Future Cost of Removal Actions

To estimate the resources that EPA will need to implement the removal program in the future, we would have to estimate both the number and the cost of the emergency responses and other removal actions that the Agency will undertake. One way to make these predictions is to consider past demand and extrapolate into the future. Alternatively, analysis of the removal planning

Table 2-4. Historical Lead Data for Removal Action Starts, FY 1992–FY 1999

Lead	Emergency responses (548)	Time-critical removal actions (1,709)	Non-time-critical removal actions (183)
Fund	98%	71%	30%
PRP	2%	29%	69%
Other	0%	0%	1%

Source: RFF removal action starts data provided to RFF by EPA, March 2000.

process could indicate future demand for removal actions. Unfortunately, neither approach is possible, as discussed below, because of a lack of data.

Past Demand

Recent historical data suggest, at least circumstantially, that EPA has faced a steady demand for removal actions. Each year over the past decade or so, EPA received about 30,000 "release notifications" that had to be evaluated for possible emergency response.[14] The main source of these notifications is the National Response Center (NRC), a call center staffed by U.S. Coast Guard personnel whose mission is to receive notifications of oil and hazardous substance releases and pass the information on to the appropriate federal response agencies for further evaluation. A large number of the releases reported to NRC are oil spills, which are addressed under the Clean Water Act (as amended by the Oil Pollution Act), or are otherwise not eligible for response under CERCLA. As Figure 2-4 illustrates, since 1996 EPA has

Figure 2-4. CERCLA-Eligible Release Notifications, FY 1992–FY 1999

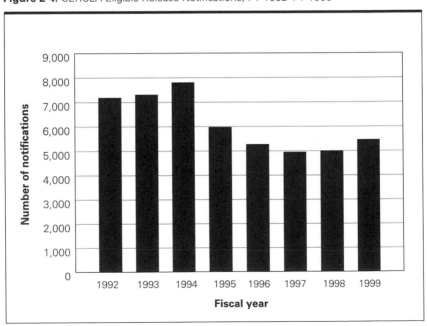

Source: Release notifications data provided to RFF by EPA, April 2000.

received approximately 5,000 notifications a year for potential responses under CERCLA authority.

State programs, however, respond to most of these releases, with EPA often providing support through technical assistance and oversight. However, since EPA does not maintain records tracking the fate of notifications within the removal program, it is not possible to determine how many become emergency responses or removal actions with EPA involvement or oversight, and for what reasons. Moreover, although the number of removal actions conducted by EPA itself has remained remarkably constant in recent years (see Figure 2-1), it is not immediately evident why. It is likely, however, that the relatively constant budget for the removal program largely accounts for the steady level of removal actions taken.

An additional piece of information that would be helpful in predicting the future cost of the removal program is the historical distribution of costs for each of the three types of removal actions. Unfortunately, EPA's databases—the Comprehensive Environmental Response, Compensation, and Liability Information System (CERCLIS) and the Integrated Financial Management System (IFMS)—do not readily permit such an analysis. IFMS does not track the cost of individual removal actions at a site; rather, costs are coded only to indicate that they are associated with removal activity in general. Thus, if multiple removal actions are taken at a site, IFMS data reveal only the total cost of all those removal actions.

By examining costs at sites at which only one removal action has been undertaken, it is possible to get a rough idea of the costs associated with conducting the three types of removal actions, even though these costs are not representative of the full universe of removal actions. Based on expenditures from FY 1992 through FY 1999, an emergency response, a time-critical removal action, and a non-time-critical removal action cost, on average, approximately $354,000, $480,000, and $662,000, respectively.[15] These averages, however, do not reflect the wide range of costs that characterize removal actions. For instance, one time-critical removal—a recent two-year response effort at the Chicago Metro MP site—cost more than $10 million, whereas some other recent time-critical removal actions have cost less than $10,000 each.[16] These wide ranges are also characteristic of emergency responses. However, even if it were possible to accurately estimate the cost of a removal action by removal type, because neither EPA headquarters nor the regions have a systematic basis for projecting the potential number and types of future removal actions, predicting future costs in this way is not feasible.

Removal Planning Process

To better understand the removal planning process within the 10 EPA regional offices, we conducted interviews with the regional program managers for the removal program. Our hope was that these interviews would enable us to identify the sources of sites referred to EPA for removal actions and to estimate the potential "pool" of such sites so that we could model the need for likely future removal actions. Our questionnaire to the EPA regions asked about the planning process for all three types of removal actions.* However, time-critical actions quickly became the focus of our discussions with the regional program managers, in large part because emergency responses are inherently difficult to predict and non-time-critical actions are less frequently taken.

The system for identifying sites at which time-critical actions might be necessary is largely informal. EPA receives notification of sites from several sources throughout the year, including state superfund programs, the federal remedial program, federal and state Resource Conservation and Recovery Act (RCRA) programs, local fire or emergency response organizations, the NRC, tribal authorities, brownfields programs, and citizen complaints. State programs, in particular those involved with hazardous waste sites and facilities, are the major sources of sites needing assistance from EPA for time-critical removal actions.

Half of the EPA regions have formal meetings with their state or federal counterparts to anticipate potential future workload and discuss priorities. These meetings range in frequency from quarterly to annually. However, regular, monthly, or even weekly informal contact by telephone to deal with emerging crises or to reorder priorities is an important part of all regions' planning processes.

None of the EPA regions had a planning horizon that exceeded one year. Even in regions with more formalized planning processes, the regional program managers stressed that planning even a few months ahead is problematic. The lead time available for a region to assess a possible time-critical removal action can be as short as a few days, which indicates that the removal program's priorities need to remain fluid as the regional program managers perform a kind of triage, trying to address the worst sites first.

The planning process for non-time-critical actions is more controlled. If these are conducted at all in a region, they are usually done under the aegis of the remedial program. Regional decision teams, consisting of managers from

---

*See Appendix D for a copy of the questionnaire.

throughout the Superfund program, exist in most regions to discuss workload planning for the sites in the remedial program as well as the need for non-time-critical and other types of removal actions.

Although all the regional program managers interviewed agreed that sites within their state systems may require removal actions, none were able to estimate the size of this pool of sites. If state programs have lists of such sites or facilities, they appear reluctant to share them. It may well be that as EPA and the states already have as many removal actions as they can handle, they do not see the need to look for more.

The most common source of sites requiring removal activities cited by the regional program managers was a category of "economically marginal" facilities, often mom-and-pop operations that cannot afford to meet regulatory requirements for handling, storage, and disposal of the hazardous materials with which they work. Examples included small electroplating shops, secondary lead smelters, and waste oil recycling facilities. In the regions' experience, a few of these operations go bankrupt on a regular basis, often leaving behind vats and drums of hazardous materials. Although the removal program or RCRA staff may know of such facilities, no formal system exists to track them.

Three regional officials expressed concern about what they believe to be thousands of abandoned or economically marginal mines in their states. Unstable tailings disposal areas and acid mine drainage threaten groundwater and surface water supplies at many of these sites. These regions undertake emergency responses and time-critical removal actions at such sites on a regular basis, but they do not have a site assessment program to catalog and assess the need for further cleanup action.

When asked about future political or economic changes that might alter the number of sites to be addressed by the removal program, few regional program managers anticipated political changes within their states that would alter the current balance. Most states continue to rely on federal programs for emergency response and removal action support, although at least one state, Pennsylvania, has a program that provides funding for implementation of emergency response programs.[17] The regional program managers did express concern about the impact of a major downturn in the economy, in which case a large number of the marginal operations—the small manufacturing facilities and mines described earlier—could fail, overwhelming the removal program with sites that need immediate attention.

Finally, the growing responsibilities of the removal program under federal statutes other than CERCLA and the new counter-terrorism initiatives, noted in Table 2-2, appear to be placing additional strain on the resources of the program. The Federal Emergency Management Agency does reimburse EPA for removal actions taken pursuant to the Stafford Act[18] in response to natural or human-caused disasters, although reimbursement is not always timely.[19] EPA, moreover, has not always received adequate funding for implementation of counter-terrorism initiatives. It is beyond the scope of this study to assess the magnitude of such additional programs, but they have clearly added to the responsibilities of the removal program staff, which has not increased commensurately.

Conclusions

Data limitations make it impossible for us to predict the future demand for federal removal capabilities or to estimate the ideal size and cost of the future removal program. Nevertheless, these questions are worth pursuing. Past data suggest a budget-driven program, in which the number and cost of removal actions remain relatively constant over time, with senior EPA management allocating additional resources when a truly huge problem arises, as was the case with methyl parathion contamination. Most regional program managers indicated that every year there are sites that they cannot address. They must constantly reorder their priorities, focusing on the most urgent problems first. Although the regions are unlikely to turn away true emergencies, they admit to taking a "band-aid" approach to some time-critical actions, doing the minimum necessary and stretching out the action over years. No data exist to document how often this approach is taken or the health and environmental implications of these delays.

Chapter Three

The Remedial Program and the Current National Priorities List

In the two decades since the Superfund statute was enacted, the U.S. Environmental Protection Agency (EPA) has cataloged more than 43,000 potentially contaminated sites in its Comprehensive Environmental Response, Compensation, and Liability Information System (CERCLIS). More than 41,000 of these sites have had a preliminary assessment to determine whether cleanup is warranted,[1] but only a small number have, in fact, been placed on the National Priorities List (NPL), which is generally considered the official register of the most hazardous sites in the nation. The NPL is the centerpiece of the Superfund program and the focus of EPA's remedial program. NPL sites addressed under the remedial program are particularly important when estimating future costs because under the Comprenesive Environmental Response, Compensation, and Liability Act of 1980 (CERCLA), it is only for these sites that EPA can fund longer-term (and more expensive) cleanups with money from the Trust Fund.

The expense and pace of cleaning up NPL sites has been, and continues to be, a contentious topic among followers of the Superfund program. In 1998, then EPA Administrator Carol Browner testified that cleanup of sites had become "faster, fairer, and more efficient."[2] Yet, even though more than half of all NPL sites have been deemed "construction-complete"—meaning that all physical remedies are in place and immediate risks posed by the site have been addressed—there remain hundreds of sites placed on the NPL during the early years of the program where cleanup remedies have still not been fully implemented. These seemingly conflicting signs contribute to confusion about the status of sites on the NPL and the amount of money EPA needs (and when the Agency needs it) to finish cleanup of these sites.

This chapter focuses on those sites listed on the NPL as of the end of FY 1999, which we refer to as "current" NPL sites. Specifically, these are the 1,245 nonfederal sites that were either listed as final on or deleted from the NPL as of September 30, 1999. We first briefly describe the authority granted to EPA by CERCLA to address NPL sites. Second, we discuss the cleanup process conducted as part of the remedial program—the succession of events undertaken at most NPL sites, referred to as the remedial pipeline. Third, we provide an overview of the cost of the remedial program, using FY 1999 expenditures as a baseline. Lastly, we outline the primary factors that drive estimates of the remaining cleanup costs for current NPL sites.

EPA's Authority to Address NPL Sites

Congress provided EPA with two mechanisms to clean up NPL sites. The first is Superfund's far-reaching liability scheme; the courts have held that liability under the Superfund statute is retroactive, strict, and joint and several. Retroactive liability means that responsible parties can be held liable for the cleanup costs attributable to generation, transportation, treatment, storage, and disposal of hazardous substances that occurred before the enactment of the statute. Strict liability means that responsible parties can be held liable for cleanup costs irrespective of negligence and even if they can show that they met all regulatory standards applicable at the time of the activity. Joint and several liability means that if harm at a site is indivisible, the government can hold one or more parties liable for the full cost of cleanup, even if other parties are liable as well.*

The Agency thus has powerful enforcement authority to issue orders or bring legal action against potentially responsible parties (PRPs). Under CERCLA,[3] a variety of parties may be held liable for cleanup costs:

- present owner(s) and/or operator(s) of the site at which the release occurs, even if they did not cause or contribute to the release;
- prior owner(s) and/or operator(s) of the site at which the release occurs, if disposal of hazardous substances occurred during their period of ownership or operation;
- generators of hazardous substances who arranged for the disposal of the substances at the site; and

---

*The burden is on responsible parties to show that the portion of the harm (and the associated cleanup cost) for which they are responsible is separable from that of other parties.

• transporters of hazardous substances if they were responsible for selecting the site.[4]

EPA's second mechanism to implement long-term cleanups is the Trust Fund. By creating this fund, Congress gave the Agency the ability to finance the cleanup of "orphan" sites—where no financially viable responsible party can be found—or sites where the responsible parties are recalcitrant. The Trust Fund allows EPA to initiate prompt cleanups, which is particularly important at sites posing imminent health hazards, for which costs can later be recovered from PRPs.

Superfund's stringent liability regime has proven extremely effective in minimizing direct costs to the government of implementing cleanups and, therefore, the costs to the Trust Fund. This has particularly been the case since 1989, when then EPA Administrator William K. Reilly announced that the Agency would pursue an "enforcement first" strategy. Under this approach, EPA has increasingly used a variety of enforcement tools to ensure that responsible parties undertake long-term cleanup efforts, known as remedial actions (RAs). For example, failure to comply with section 106(a) of CERCLA can result in fines of $25,000 a day.[5] In addition, failure to provide for removal or remedial action pursuant to an order issued under section 106(a) of CERCLA can also result in treble damages,[6] although this tool has rarely been used. The enforcement-first policy has had a dramatic impact. During the first 10 years of the program, EPA used Trust Fund money to pay for well over half of all RAs. Since the enforcement-first policy took hold in FY 1991, EPA has financed a considerably smaller percentage of RAs than have PRPs; from FY 1991 through FY 1999, EPA paid for only 27% of all RAs.[7]

The Remedial Cleanup Process

Under CERCLA, EPA has the authority to conduct two types of cleanup— removal actions and remedial actions. As discussed in Chapter 2, removal actions are relatively short-term and are typically emergency measures designed to eliminate the threat of imminent hazards (such as the removal of leaking barrels or the cleanup of spills or other types of surface contamination). Most often, the goal of a removal action is to address immediate risks to human health. Removal actions paid for by EPA (which most are) are limited by law to $2 million and a duration of one year (unless a waiver is issued) and can be implemented whether or not a site is on the NPL.[8]

Remedial actions, in contrast, generally take longer and are more expensive and more permanent; they are part of a sequence of events that constitute the Superfund pipeline. We limit our discussion here to the remedial phases of the pipeline—the remedial investigation/feasibility study (RI/FS), the remedial design (RD), and the RA. These phases are the heart of the remedial program and constitute the cleanup process most often used to address NPL sites.

The remedial pipeline begins with an RI/FS. The remedial investigation (RI) and the feasibility study (FS) are usually conducted concurrently, although they can also be performed sequentially. The RI involves the collection of data to characterize site conditions, determine the concentration level of contaminants, and assess the risk posed to human health and the environment. The FS focuses on the development, screening, and initial evaluation of remedial alternatives; it generally establishes remedial action objectives that specify contaminants of concern, potential exposure pathways, and remediation goals. These studies can be carried out by EPA, in which case they are referred to as Fund-lead (often performed by contractors, the U.S. Army Corps of Engineers, or states using EPA funds), or by PRPs, referred to as PRP-lead.

In the final part of the RI/FS phase, EPA issues a proposed plan setting forth the preferred remedial alternative. After public comment on the proposal (usually a 30- to 60-day period), EPA makes a final decision in a formal document called a record of decision (ROD), which marks the completion of the RI/FS. Remedy selection is based on two threshold criteria—overall protection of human health and the environment, and compliance with applicable or relevant and appropriate requirements (ARARs).* Remedies also must balance such factors as long-term effectiveness and permanence; reduction of toxicity, mobility, or volume through treatment; short-term effectiveness; implementability; and cost.[9] The selection of a remedy is solely the province of EPA and may not be delegated to PRPs; EPA signs all RODs, making the Agency the final arbiter of cleanup decisions at NPL sites.

The next phase of the remedial process is the RD, when the engineering plan used to guide implementation of the remedy is developed. The RA is the final phase of the process, when the remedy is put in place and actual cleanup

---

*ARARs refer to federal and state environmental laws, regulations, and cleanup standards that specifically address a hazardous substance, pollutant, or contaminant, remedial action, location, or other circumstances at a Superfund site. CERCLA sec. 121(d)(2)(A)(i) and (ii).

at the site begins. The remedy constructed at the site must correspond to the remedy indicated in the ROD, although if new information emerges that suggests a change should be made, the remedy can be modified through the issuance of a ROD amendment or an "explanation of significant difference."

In reality, sites rarely move through the remedial pipeline in the linear, step-by-step manner described above. Most sites are divided early in the cleanup process into multiple projects, known as operable units, that typically separate the site geographically, by media (e.g., soil, groundwater), or by remedy (e.g., landfill cap, groundwater restoration). Remedial activity at sites with multiple operable units is generally staggered. The RI/FS at one operable unit, for example, may start a year or more after the RI/FS begins at another. Operable units may also move through the pipeline at different paces because of a number of factors, such as the availability of funding, the complexity of the cleanup, or the level of cooperation of responsible parties. There also can be multiple actions of the same type (e.g., two RAs) at a single operable unit. The discovery of new information can even push an operable unit backward to an earlier stage of the remedial pipeline.

Typically, actions at a single operable unit overlap. RAs, for instance, generally begin prior to the completion of the RD. Also, because EPA increasingly uses phased funding, actions at multiple stages of the pipeline may occur simultaneously at a single operable unit. Thus, the remedial pipeline should be thought of as a general guide to the remedial cleanup process, rather than as an exact characterization of how the process actually unfolds at all sites and all operable units.

The three phases—RI/FS, RD, and RA—form the core of the remedial pipeline, but there are a number of other site-specific activities that are integral to the remedial cleanup process. These include oversight of PRP-lead remedial pipeline actions, enforcement efforts (by both EPA and the Department of Justice), response support, and community and state involvement, as well as management and support activities. All of these activities are essential components of the process, and EPA devotes considerable resources to them. In addition, when a remedy is complete, operation and maintenance activities are often required to ensure that the remedy continues to function as designed. As discussed in Chapter 4, some operation and maintenance activities include actual cleanup. For example, long-term response action, or LTRA, often involves the long-term operation of pump-and-treat systems to restore groundwater. For this reason, LTRA can also be characterized as part of the remedial pipeline (as we do in Chapter 7).

Cost of the Remedial Program

The largest component of Superfund expenditures is, as one would expect, the cost of the remedial program. In FY 1999 a total of $622.3 million, or 40% of Superfund program expenditures, went to remedial activities *directly* charged to sites. The overwhelming majority (95%) of these costs went to activities at NPL sites.* The remaining site-specific costs were incurred at non-NPL sites, which are discussed later in this chapter.

The majority of remedial program costs are for conducting Fund-lead actions at NPL sites. Expenditures for these actions in FY 1999 were $478.1 million. EPA spent an additional $114.8 million to carry out other site-specific activities, bringing the total expenditures charged directly to NPL sites in FY 1999 to $592.9 million. These other site-specific activities are generally equivalent to those performed in support of removal actions and were defined in Chapter 2. (A few additional site-specific activities, such as the provision of technical assistance grants, are carried out only in support of remedial clean-ups.) The additional $114.8 million in site costs included $33.2 million for oversight of PRP-lead actions, $55.8 million for enforcement efforts (of which $30.5 million was spent by EPA and $25.3 million by the Department of Justice), $15.3 million for response support activities, just under $10 million for community and state involvement activities, and about $0.5 million for five-year reviews, which are the evaluations of site cleanup efforts that EPA conducts five years after the first remedy at a site is initiated (discussed in detail in Chapter 4).

An additional $29.4 million was spent in FY 1999 on remedial activity at non-NPL sites. Of this total, $12.3 million was spent for Fund-lead actions (such as RI/FSs and RDs), and the remaining $17.1 million went to oversight, enforcement, response support, and community and state involvement.

The true cost of the remedial program is even larger than the $622.3 million that is charged directly to NPL sites. The cost of developing national policy and guidance documents and other activities performed at EPA headquarters and in the 10 regional offices, for instance, are not charged directly to sites but contribute to the overall remedial program. However, because of the way expenditures are coded in EPA's Integrated Financial Management System (IFMS), it is difficult to cull the proportion of such costs that should be directly attributed to the remedial program. We discuss these activities and their associated costs in Chapter 6.

---

*Included in this total is about $21.8 million spent at sites proposed to the NPL. More than three-quarters of these expenditures were for RI/FS expenses.

Determinants of Cleanup Costs

To estimate the cost to EPA of cleaning up NPL sites over time, some basic questions must be answered. What types of sites are on the NPL? How many remedial pipeline actions will there be? How many actions will EPA pay for with Trust Fund dollars? How much will these actions cost? When will EPA need the money to pay for them? The answers to such questions are central to estimating future Superfund costs and form the building blocks for the model we use to estimate the cost to EPA of cleaning up sites on the NPL from FY 2000 through FY 2009. In the sections below, we describe each of these building blocks.

Categories and Types of Sites

Sites on the NPL range from abandoned mines to currently operating industrial facilities to fields where hazardous substances were dumped. NPL sites also vary in terms of the contaminants of concern, the size of the site, and the total cost of cleanup. As a result, there are many possible ways to categorize NPL sites.

At the end of FY 1999, there were 1,245 nonfederal sites on the NPL: 1,055 final NPL sites and 190 deleted NPL sites.[10] An additional 52 sites had been proposed for listing. The status of an NPL site serves as a rough indicator of how much cleanup remains. Proposed sites include both sites that were recently proposed to the NPL and sites that were proposed years ago but for some reason have not been (and may never be) made final on the NPL. Because cleanup activity is often at an early stage at proposed sites, there generally is considerable work still to be done. At the other end of the spectrum are deleted sites at which, by definition, all remedial actions have been completed (or, in a few cases, were not required because EPA decided the site did not warrant cleanup or had been sufficiently addressed with removal actions). The rest of the sites are listed as final on the NPL and can be at any stage of the remedial cleanup process.

Whether a site is "construction-complete" is another way NPL sites are typically categorized. Construction completion is a designation established by EPA in 1990 to characterize sites where "physical construction of all cleanup actions are complete, all immediate threats have been addressed, and all long-term threats are under control."[11] At the end of FY 1999, 648 nonfederal NPL sites had been designated by EPA as construction-complete, leaving a balance of 597 sites that were not.

Among the sites that are not yet construction-complete, 318 have been on the NPL for a long time. Colloquially, EPA refers to these sites as "teenagers" because they were proposed to the NPL prior to FY 1987—that is, they have been on the NPL for at least 13 years. It is not clear why the remedial cleanup process is taking longer at these sites than at other NPL sites, although many of the most expensive and largest sites are among them. One explanation may be that during the early years of the Superfund program, a large number of sites were listed as final on the NPL—707 of the current 1,245 sites were added to the NPL from FY 1983 through FY 1986[12]—and that this sudden influx of sites overwhelmed the Agency's resources and its capacity to address all sites in a timely manner. Interestingly, a large percentage of teenager sites are clustered in two EPA regions, Region 2 and Region 5, which have 29% and 22% of these sites, respectively.

Since the focus of this study is cost, we pay special attention to NPL sites with actual or expected total removal and remedial action costs of $50 million or more, sites we refer to as mega sites.[13] Sites with both predominantly Fund-lead and PRP-lead remedial pipeline actions are included in this definition. EPA identified 112 sites on the NPL at the end of FY 1999 that met the $50 million threshold—109 final sites and 3 deleted sites. EPA also identified 7 mega sites that were proposed to the NPL at the end of FY 1999.*

Table 3-1 shows how the 1,245 nonfederal final and deleted sites on the NPL at the end of FY 1999 fit into the categories discussed above. We use these categories of sites in developing our model of future costs.

Sites can also be categorized by the type of industrial facility or activity responsible for the contamination. This approach has proven useful for the purpose of estimating future cleanup costs.[14] We classify each site on the NPL at the end of FY 1999 as one of 11 site types.† Figure 3-1 displays the number of sites in these categories.

Non-NPL Sites

Before discussing other factors that are important to consider when estimating future costs, it is worth mentioning that EPA also conducts site studies and addresses contamination at sites that are *not* listed as final on the NPL. There are four main categories of non-NPL sites:

---

*See Appendix D for a list of these sites.

†See Appendix F for a description of each site type.

Table 3-1. Status and Category of NPL Sites, End of FY 1999

1,245 Final and deleted NPL sites		
112 Mega sites		
	25	Construction-complete
	87	Not construction-complete
		68 Teenager
		19 Nonteenager
1,133 Nonmega sites		
	623	Construction-complete
	510	Not construction-complete
		250 Teenager
		260 Nonteenager

Source: RFF site-level dataset; does not include federal facilities.

- sites that have removal actions but are not subject to longer-term cleanups (discussed in Chapter 2);
- sites proposed to, but not final on, the NPL;
- sites referred to as NPL equivalent sites; and
- other sites, not included in any of the three prior categories, at which EPA conducts site studies or enters into enforcement agreements with PRPs to address contamination at the site.

Generally, most sites have become final shortly after being proposed for NPL listing, but this is not necessarily the case. Of the 52 sites on the NPL at the end of FY 1999 categorized as proposed sites, 21 had been proposed for listing for longer than five years, suggesting that they were unlikely to be made final NPL sites. The distinction between proposed and final NPL sites is that EPA may use Trust Fund dollars *only* for RAs at sites that are final on the NPL.[15] At its discretion, however, the Agency can fund other types of actions (such as RI/FSs and RDs) at any site, and it routinely does so at proposed NPL sites and at other sites that are not—and may never be—proposed to the NPL.

In FY 1999, 67% ($16.8 million) of the total spent at proposed NPL sites was for RI/FSs. EPA often funds RI/FSs at proposed sites to speed up the remedial process. On an individual basis, however, these sites will not require considerable EPA resources unless they become final NPL sites.

Figure 3-1. NPL Sites by Site Type

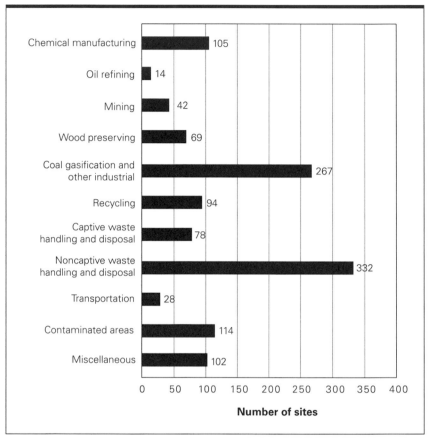

Source: RFF site-level dataset. Based on 1,245 nonfederal final and deleted NPL sites as of the end of FY 1999.

NPL equivalent sites are another type of non-NPL site that warrants discussion. EPA defines an NPL equivalent site as one at which responsible parties perform cleanup under EPA enforcement authority and with EPA oversight, but without being listed as final (or sometimes even proposed) on the NPL. NPL equivalent sites are predominantly—if not completely—PRP-lead, where EPA has been able to negotiate a legal agreement with the responsible parties governing cleanup, typically with the support of the relevant state agency. The sites are considered NPL equivalent because they are eligible for NPL listing (i.e., they have a hazard ranking system score of 28.5 or above) and the response action will, according to EPA, be carried out in a manner consistent with the National Contingency Plan and applicable Agency guidance for NPL sites.[16]

Of the 52 sites that were proposed to the NPL at the end of FY 1999, as of the end of FY 2000, 16 were NPL equivalent sites, 16 became NPL final sites during FY 2000 (1 site was later ordered off the NPL as the result of a court decision), and 15 sites still had listing decisions pending (5 of which were under state agency lead). The remaining sites included 2 NPL equivalent candidate sites, 2 sites deferred to a state program, and 1 site deferred for action under the Resource Conservation and Recovery Act program. Thus, of the 37 proposed NPL sites that were not made final in FY 2000, most are likely never to become final NPL sites. In fact, all but 3 of the 16 NPL equivalent sites were proposed to the NPL at least five years ago.*

Many senior EPA officials believe that the NPL equivalent approach will result in reduced negotiation time and save federal funds by increasing the number of PRP-lead remedial cleanup actions. The Agency also expects a reduction in both federal and PRP transaction costs associated with negotiating and entering into settlement agreements, as these agreements are generally developed earlier in the process than is the case at a typical NPL site. Finally, at the majority of these sites, there will be no need for cost recovery because PRPs will be conducting all response actions under a settlement agreement and will also pay EPA's future oversight costs.[17]

Although the NPL equivalent approach is an attractive alternative to listing sites on the NPL, without further analysis it is premature to conclude that NPL equivalent cleanups (and public involvement efforts) are in fact *equivalent* to NPL cleanups. Additionally, it is not clear whether the NPL equivalent approach is extending the reach of the Superfund program, or merely changing the "status" of sites. It may be that the sites EPA is now addressing as NPL equivalent sites would have previously been listed on the NPL and addressed in a similar manner—that is, as PRP-lead sites.

At the end of FY 1999, EPA had a workload of about 70 NPL equivalent sites. Among them were 8 mega sites, including the General Electric–Housatonic River site, a large contaminated sediments site in Massachusetts that was proposed to the NPL in September 1997. Currently, EPA does not dedicate a large portion of its budget to NPL equivalent sites. In FY 1999, the Agency spent less than $12 million for site-specific activities (e.g., remedial response, oversight, enforcement) at these sites. Moreover, based on interviews with the 10 regional

---

*See Appendix B for the proposal date for each of the 52 proposed NPL sites, as well as their status at the end of FY 2000.

Superfund division directors, it does not appear that the number of NPL equivalent sites is likely to increase appreciably in the future, although some regional offices are more interested in pursuing this approach than others.

Finally, as noted earlier, there are a host of other non-NPL sites where EPA may conduct site studies (such as RI/FSs) or bring enforcement action against or negotiate agreements with PRPs for cleanups. CERCLA liability reaches, at least in theory if not in practice, all sites in the nation. Thus, from the inception of the Superfund program, EPA has always carried out some activities at sites not on the NPL. For instance, in FY 1999, EPA spent $2 million on RI/FSs at these sites (coded "ZZ" in IFMS), and another $11.3 million on other activities, mostly general response support.

Number of Fund-Lead Actions

Remedial pipeline actions are generally distinguished by whether they are Fund-lead (i.e., managed and paid for by EPA; most actions are actually conducted by contractors or the U.S. Army Corps of Engineers) or PRP-lead (i.e., financed by a responsible party or parties).* There are, in fact, very few sites at which *all* response actions are either Fund-lead or PRP-lead. For example, at perhaps the most notorious NPL site, Love Canal, five operable units had Fund-lead RAs and one operable unit had a PRP-lead RA. It is also common for the lead of actions to change during or between phases of the remedial pipeline. The majority of RI/FSs are Fund-lead, even at operable units that will eventually have PRP-lead RDs and RAs. By funding the RI/FS, EPA can expedite the cleanup process while it identifies and negotiates a settlement with the PRPs, who often ultimately conduct and finance the RD and RA. Changes in lead occur most often between the RI/FS phase and the RD phase. Table 3-2 illustrates how the ratio of Fund-lead to PRP-lead actions changes through the pipeline, as well as how it has changed over major phases of the Superfund program's history. Table 3-2 reflects two key dates in the development of the Superfund program: FY 1986, when the Superfund Amendments and Reauthorization Act was passed, and after FY 1990, when the enforcement-first policy began to take effect.

Historically, EPA has performed a higher percentage of RI/FSs than have PRPs. However, as noted above, this lead ratio switches as operable units move into the RD and RA phases of the pipeline. From FY 1991 through FY

---

*In actuality, there are a variety of leads reflecting the many different funding and work arrangements. See Appendix F for an explanation of how we organized leads in our analyses.

Table 3-2. Comparison of Leads for Remedial Pipeline Actions by Time Period (Percentage)

Period	RI/FS		Remedial design		Remedial action	
	Fund-lead	**PRP-lead**	**Fund-lead**	**PRP-lead**	**Fund-lead**	**PRP-lead**
FY 1980–FY 1986	76%	24%	63%	37%	67%	33%
FY 1987–FY 1990	52%	48%	49%	51%	54%	46%
FY 1991–FY 1999	54%	46%	28%	72%	27%	73%

Source: Data provided to RFF by EPA, October 2000.

1999, there have been almost three PRP-lead RAs for every one Fund-lead RA, attesting to the success of the enforcement program.[18]

Interestingly, the percentage of Fund-lead and PRP-lead actions varies significantly among site types. For example, at recycling sites and oil refining sites, from the time that the enforcement-first policy took hold in FY 1991 through FY 1999, the percentage of PRP-lead RAs was 81% and 80%, respectively. In contrast, during the same period, only 39% of RAs at mining sites were PRP-lead.*

The number of Fund-lead actions, and particularly the number of Fund-lead RAs, largely determines the amount of money EPA will need for cleanup costs at NPL sites. The number of Fund-lead RAs is also critically important to states, since they are responsible for 10% of the costs of Fund-lead RAs undertaken at sites within their borders, and 100% of the costs of any operation and maintenance activities that these remedies may require.[19] The future number of PRP-lead actions is also important for estimating future costs because EPA and the states use Trust Fund resources to oversee these actions. Most sites on the current NPL have progressed at least partially through the pipeline, which makes it possible to predict the leads of future actions with some degree of confidence.

Unit Costs of Fund-Lead Actions

The single largest financial responsibility for EPA at NPL sites is paying for Fund-lead actions. In FY 1999, nearly a third of the Agency's total Superfund

---

*See Appendix F for further discussion about leads and a historical breakdown of action leads by site type.

expenditures went toward Fund-lead RI/FSs, RDs, and RAs. Including both extramural (e.g., contractors) and intramural (e.g., staff time) costs, EPA spent nearly $480 million on these actions in that year.

Developing cost estimates for future Fund-lead actions is complicated. EPA has not historically kept track of its own expenditures on an action-by-action basis at NPL sites, so the Agency has no simple way of knowing how much it has spent on individual actions at individual operable units. All agree, however, that the lion's share of the money has been spent on Fund-lead RAs.

Most prior studies of Superfund costs have used data from ROD documents to estimate likely RA costs.[20] For each remedy under consideration, EPA conducts an analysis of the likely capital and operation and maintenance costs. These estimates are made prior to the design of the remedy, let alone its implementation, and serve only as general indicators of the real cost of implementing the remedy.

To our knowledge, no one has used actual expenditure data to estimate the cost of individual actions at NPL sites, the approach taken here.[21] EPA provided us with expenditure data from its internal financial management system, IFMS, which we use to create the RFF IFMS dataset.* Expenditure data in the RFF IFMS dataset includes two major types of expenditures—those charged directly to specific sites and those charged to other program activities. Each site-specific expenditure in IFMS is linked to a particular site and to either a remedial pipeline phase—RI/FS, RD, or RA—or other common cost category (e.g., oversight, enforcement, response support). Moreover, each expenditure is coded extramural (contract, cooperative agreement with state, or interagency transfer) or intramural (Department of Justice, payroll, or travel and supplies). Although Department of Justice expenditures are made external to EPA, we classify them as intramural because Department of Justice staff are carrying out site-specific enforcement efforts.

We develop unit costs for Fund-lead pipeline actions—RI/FSs, RDs, and RAs—by using data from IFMS in concert with data from CERCLIS.† In brief, to calculate extramural costs, we total the number of completed and partial Fund-lead RI/FSs, RDs, and RAs conducted from FY 1992 through FY 1999‡ and divide the sum into the total extramural dollars (in constant 1999

---

*See Appendix C for an explanation of the RFF IFMS dataset.

†See Appendix F for a detailed description of the methodology we use to calculate unit costs.

‡We chose this time period because it reflects a mature Superfund program and it provides us with sufficient data for analysis.

Table 3-3. Unit Costs of Remedial Pipeline Actions (at Operable Unit Level) by Site Category (1999$)

	Type of action		
	RI/FS	**Remedial design**	**Remedial action**
Extramural unit costs			
All sites	1,300,000	1,300,000	11,000,000
Mega sites	2,500,000	3,900,000	30,000,000
Nonmega sites	990,000	880,000	5,800,000
Intramural unit costs			
All sites	63,000	31,000	59,000
Mega sites	82,000	50,000	120,000
Nonmega sites	57,000	27,000	43,000

dollars) spent for these actions over this same period. We apply the same methodology to calculate intramural unit costs but use data only from FY 1996 through FY 1999.*

The results, presented in Table 3-3, show the unit cost for each of the three main remedial pipeline actions, which we calculate for actions conducted at both mega sites and nonmega sites. These unit costs represent the average dollars spent for Fund-lead RI/FSs, RDs, and RAs at the operable unit level. Not surprisingly, whether the action takes place at a mega site or a nonmega site makes a considerable difference. It is important to emphasize that these unit costs represent only the costs of implementing and managing Fund-lead RI/FSs, RDs, and RAs. These unit costs do not cover other site-specific activities, such as community and state involvement, nor do they include other program costs necessary to implement the remedial program, such as rent. Thus, they are not "fully loaded" costs.

To serve as a check on our unit costs, we conducted a survey of remedial project managers (RPMs) and created the RFF remedial action cost database. For 158 Fund-lead RDs completed from FY 1995 through FY 1999, we asked each RPM to indicate the cost of the subsequent RA in terms of the value of the accepted RA contract bid—in other words, the amount of money EPA

---

*As we explain in Appendix F, intramural expenditure data prior to FY 1996 are problematic.

was to pay the contractor to implement the RA. When feasible, the RPMs segregated RA costs from other activities included in the contract (e.g., operation and maintenance). The average RA contract cost was approximately $29.5 million for mega sites and $5.4 million for nonmega sites; both figures are remarkably close to our extramural unit cost estimates.*

We also calculated average extramural and intramural costs for RI/FSs, RDs, and RAs for each of the 11 site types (based solely on actions conducted at non-mega sites) described earlier in this chapter. Although there is some minor variation in intramural unit costs among the site types, there is far more variation in extramural costs. As shown in Table 3-4, for nonmega sites the extramural unit cost for one operable unit for an RI/FS range from $440,000 for a captive waste handling and disposal site to $2.2 million for a mining site. Average RD costs range from $450,000 for a transportation site to $1.2 million for a mining site or a noncaptive waste handling and disposal site. RAs are much more expensive and range from a low of $3.2 million at miscellaneous sites (such as agricultural, dry-cleaning, and military-related facilities) to a high of $11 million at a wood preserving site. The most expensive sites to clean up are mining sites, wood preserving sites, and transportation sites, all of which have average costs of more than $10 million for an RA at a single operable unit.

EPA recently reported that the cost of cleanup has been reduced by approximately 20% because of the 1995 administrative reforms regarding remedy selection. The Agency points to a number of factors responsible for these cost savings, including the use of presumptive remedies,[22] the consideration of reasonably anticipated future land use determinations that allow cleanups to be tailored to specific sites, and the use of a phased approach to defining objectives and methods for groundwater cleanups.[23] These cost savings relate only to the extramural dollars associated with Fund-lead RAs and are not expected to carry over to other remedial pipeline actions or to intramural costs.

Our evaluation of Fund-lead actions, however, suggests that cleanup costs may not, in fact, be declining. Although there are too few observations to allow definitive conclusions, our analysis shows that the cost of Fund-lead RAs has actually been *increasing* in recent years. The average cost of an RA (at the operable unit level) completed during FY 1992 through FY 1995 was $10.2 million, but the average cost of an RA completed during FY 1996 through FY 1999 was $11.9 million, according to IFMS expenditure data.† Since EPA

---

*See Appendix D for further details on the RFF remedial action cost database.

†These average costs are calculated for RAs conducted at all sites. It is worth noting that the average RA cost increased at both mega sites and nonmega sites.

Table 3-4. Extramural Unit Costs of Remedial Pipeline Actions (at Operable Unit Level) at Nonmega Sites by Site Type (1999$)

Site type	Type of action		
	RI/FS	Remedial design	Remedial action
Chemical manufacturing	710,000	830,000	5,200,000
Oil refining	550,000	910,000	4,500,000
Mining	2,200,000	1,200,000	10,000,000
Wood preserving	750,000	1,000,000	11,000,000
Coal gasification and other industrial	1,100,000	810,000	5,700,000
Recycling	1,200,000	1,100,000	4,400,000
Captive waste handling and disposal	440,000	780,000	3,700,000
Noncaptive waste handling and disposal	1,000,000	1,200,000	5,900,000
Transportation	560,000	450,000	10,000,000
Contaminated areas	970,000	650,000	4,100,000
Miscellaneous	1,200,000	610,000	3,200,000

based projected cost savings on estimates in recently signed RODs, it may be that any savings from the factors enumerated above have not yet materialized and will not show up until these more recently selected remedies are implemented. Or it may be that the cost data in RODs do not reflect the actual cost of implementing RAs, or that the RODs EPA considered are not representative of the full range of ROD estimates.*

Duration of Remedial Pipeline Actions

The pace of cleanup at Superfund sites has long been a contentious issue. According to EPA officials, there has been a steady rise in the number of construction completions, and "Superfund cleanup durations have been reduced approximately 20%, or two years on the average."[24] Whether the rise in the number of construction-complete sites is due to a quickened cleanup pace is

---

*We found that mega sites appear to be underrepresented in EPA's ROD cost database.

questionable, however. A more probable explanation is simply that more sites have reached the cleanup stage of the pipeline than in earlier years of the program—what is referred to as the pipeline effect.[25] Moreover, the methodology EPA generally uses for calculating average durations does not provide a complete view of the pace of cleanup.

Average durations for remedial pipeline actions conducted from FY 1993 through FY 1999, as calculated by EPA, are summarized in Table 3-5. These data suggest that cleanup at a typical operable unit will take a little more than eight years—the time from the date the site is proposed to the NPL until the remedial action is completed. However, in estimating the duration of pipeline actions, EPA includes only those actions that have been completed. Thus, actions that are started but not finished during the period—likely including the lengthiest actions—are excluded from Agency calculations. Some remedies, such as monitored natural attenuation, bioremediation, and soil vapor extraction, require decades of operation to achieve cleanup goals. Exclusion of these "long RAs" results in a shorter average duration.

Another aspect of EPA's average durations that can be misleading and suggest shorter average durations is that they are calculated on an action rather than on an operable unit basis. For instance, if an operable unit has two RI/FSs, EPA considers each action separately. Although legitimate for estimating the average length of individual actions, this approach does not accurately reflect how long it takes for a specific phase of the remedial pipeline to be completed. EPA's durations are often used as estimates of the length of each remedial pipeline stage, which results in an underestimate of the true amount of time it takes for an operable unit to complete a pipeline phase.[26]

Table 3-5. EPA Durations of Remedial Pipeline Actions

Pipeline phase	Number of observations	Average duration (years)
NPL proposal date to RI/FS start	164	1.0
RI/FS	117	2.6
RI/FS completion to remedial design start	440	1.0
Remedial design	496	1.7
Remedial action	409	1.8
Total		**8.1**

Source: Data provided to RFF by EPA, May 2000. Based on completed actions, FY 1993–FY 1999.

EPA's durations are also often misinterpreted as showing the total length of time it takes to clean up an entire NPL site. In fact, these durations (notwithstanding the problems noted above) represent the time for cleanup of one operable unit and should not be used to estimate the total length of the remedial pipeline at sites with multiple operable units. As noted earlier, operable units at sites generally do not move through the pipeline on a parallel track, and more than 40% of the sites on the NPL have more than one operable unit, as shown in Table 3-6. In fact, at sites with more than one operable unit, the second operable unit typically does not enter the remedial pipeline until about a year after the first operable unit.*

To provide a more accurate picture of the pace of cleanup, we calculate average durations differently. First, we consider all actions that started in FY 1992 or later,† whether or not they have been completed. To estimate the duration of actions not yet completed, we use the action's planned completion date, which is assigned by the RPM overseeing the action. Second, we consider each pipeline action at the operable unit level. That is, we collapse multiple actions of the same type into a single action for the purposes of calculating the duration of each operable unit.‡ In other words, where an operable unit had, for example, two RI/FSs, we use the earliest start date and the latest actual or planned completion date of the two RI/FSs as the start and end dates of the RI/FS phase. In this way, our estimates account for the actual time it takes an operable unit to move through each phase of the remedial pipeline.

Lastly, to better understand the effect of long RAs, we calculate the average duration of RAs in two ways: including and excluding these actions. Long RAs typically include bioremediation or soil vapor extraction as part of the remedy. These remedies, even if performing as designed, generally take decades of operation to achieve the cleanup objectives set forth in the ROD. As discussed in Chapter 7 and detailed in Appendix G, when we use average durations to forecast when actions will be conducted in our model (and when EPA will need funds to pay for them), we treat these long RAs separately. We found that about 4% of the RAs started in FY 1992 or later were expected to have durations of 10 years or more,** with an average of 20.4 years.

---

*See Appendix G for further discussion of this lag time.

†Again, we chose this date because it reflects the beginning of a mature Superfund program and it provides us with sufficient data for analysis.

‡We "collapse" actions by assigning the new, collapsed action the start date of the earliest uncollapsed action and the completion date of the last uncollapsed action.

**We consider a remedial action with a duration of 10 years or more to be a long RA.

Table 3-6. Operable Units per NPL Site

Operable units per site	Number of sites	Percentage
0	16[a]	1.3%
1	697	56.0%
2	325	26.1%
3	110	8.8%
4	46	3.7%
5	17	1.4%
6	18	1.4%
7	4	0.3%
8	4	0.3%
9	1	0.1%
10 or more	7	0.6%
Total	**1,245**	**100.0**%

[a]Nearly all of these sites were listed in the early years of the Superfund program and were never divided into operable units. Most have since been deleted from the NPL. One site was listed in FY 1999, and as of the end of that year, the number of operable units at the site had yet to be determined.

Source: RFF site-level dataset. Based on 1,245 nonfederal final and deleted NPL sites as of the end of FY 1999.

Figure 3-2 shows the results of our duration analysis compared with that of EPA.* For each of the major remedial cleanup phases—RI/FS, RD, and RA— our durations exceed the Agency's by a considerable margin. This is most evident with respect to RAs. Our analysis shows that when long RAs are included, the duration of an average RA is more than double what EPA data suggest.

We also examined whether the durations of remedial pipeline actions vary by such factors as site type and lead, and whether the action is conducted (or being conducted) at a mega or nonmega site, teenager or nonteenager site. With respect to site type, our analyses showed some minor variations for each stage of the pipeline, but the data do not suggest a discernible pattern.

There also is minimal difference in the average length of pipeline phases for Fund-lead and PRP-lead actions. The most notable difference is the period

---

*See Appendix F for a description of the methodology we use to calculate average durations.

Figure 3-2. Comparison of RFF and EPA Durations of Remedial Pipeline Phases

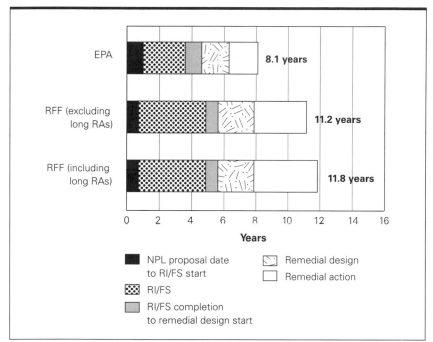

Source: EPA durations provided to RFF by EPA, May 2000, based on completed actions, FY 1993–FY 1999. RFF durations are measured at the operable unit level and are based on actions starting in or after FY 1992, using actual and planned completion dates.

from NPL proposal to the start of the first RI/FS. When the first RI/FS at a site is Fund-lead, it usually begins about six months after the site is proposed to the NPL. Since EPA is permitted to spend money on non-NPL sites, some of these RI/FSs may have begun before the proposal of the site. When the first RI/FS is PRP-lead, it usually begins about 14 months after the site is proposed to the NPL. The difference is not surprising, since it takes time for EPA and PRPs to negotiate an agreement before beginning an RI/FS.

There are also only small differences between the average durations of actions conducted (or being conducted) at mega sites and nonmega sites. Although an operable unit at a mega site takes on average about a year and a half longer than an operable unit at a nonmega site to move completely through the pipeline, more than a year of this difference is for the RI/FS stage. The rest of the actions generally take about the same time.

Table 3-7. Durations of Remedial Pipeline Phases (Years)

Pipeline phase	Mega sites		Nonmega sites	
	Teenager	Nonteenager	Teenager	Nonteenager
RI/FS	5.0	5.0	4.9	3.6
RI/FS completion to remedial design start	0.8	0.3	1.0	0.7
Remedial design	2.1	2.0	2.6	2.1
Remedial action	4.2	3.2	3.8	3.0
Total	**12.0**	**10.6**	**12.4**	**9.4**

Note: Based on actions at nonfederal NPL sites starting in or after FY 1992, using actual and planned completion dates. RA durations do not include long RAs. Totals may not add exactly because of rounding.

Lastly, we compared the durations of remedial pipeline actions conducted (or being conducted) at teenager sites—those 318 sites that were listed on the NPL before FY 1987 and were still not construction-complete by the end of FY 1999—and nonteenager sites. More than any of the other factors that we examined, the age of the site has significant impact on its duration. At each stage of the pipeline, the duration of actions at teenager sites exceeds that at nonteenager sites. As illustrated in Table 3-7, except for RI/FS durations, this is true for both mega sites and nonmega sites.

To capture these differences in average durations in our estimates of future costs, we use the average durations shown in Table 3-7, rounded to the nearest quarter, to project when future actions will take place at nonmega sites already on the NPL at the end of FY 1999. For future actions at mega sites, we use site-specific duration information gathered from staff in the 10 EPA regional offices.* For future actions at nonmega sites on the current NPL for which we do not have EPA planning data, we use different average durations for teenager and nonteenager sites. For actions to be conducted at sites added to the NPL in FY 2000 and after, we use average mega and nonmega site durations calculated from actions conducted (or being conducted) only at nonteenager sites. Although there could be future teenager sites, the slowed pace of listings in recent years would suggest that this is not likely.†

---

*See Appendix D for more information on the RFF mega site survey.

†See Appendix G for further detail on how these durations are used in our model.

The "Current" Current NPL

Since the end of FY 1999 (the end point of our data), the composition of the NPL has changed as sites have been added and deleted. In FY 2000, 36 new sites were added to the NPL,* bringing the total number of nonfederal NPL sites to 1,281, of which 1,069 were final and 212 were deleted. Among the sites listed in FY 2000, two are expected to be mega sites: Leviathan Mine in California (Region 9) and Midnite Mine in Washington (Region 10). Of the 1,281 sites, 728 sites were deemed by EPA to be construction-complete, leaving a balance of 553 sites. At the end of FY 2000, there were also 53 sites proposed to the NPL, including three expected mega sites proposed in FY 2000: Molycorp, Inc., in New Mexico (Region 6), Malone Service Company, Inc., in Texas (Region 6), and Portland Harbor in Oregon (Region 10), the last of which became a final NPL site during the first quarter of FY 2001.

Conclusions

The remedial cleanup process EPA carries out at NPL sites is the most visible aspect of the Superfund program, and the one that has garnered the most attention. Our analysis of the costs and duration of the phases of the remedial pipeline suggests that there is little evidence to date to directly support EPA's claims that the cleanup of NPL sites has become cheaper and faster. Recently, even officials at the Agency have recognized that remaining site cleanups are increasingly complex, which may force a slowdown in the number of sites that become construction-complete each year.[27]

Although every site on the NPL is unique and the remedial cleanup process plays out differently at each, NPL sites can be grouped into a number of categories, such as site type, mega or nonmega site, and teenager or nonteenager site. Each of these categories has implications for the cost and duration of cleanup, as well as the likelihood that it will be Fund-lead.

Assumptions about the duration of cleanup actions, as well as their cost and the proportion that will be Fund-lead, are critical inputs to any calculation of the future cost of addressing sites currently on the NPL. Durations determine not only when cleanup actions will be completed but also the

---

*Technically, 37 sites were added to the NPL in FY 2000, but one, Georgia-Pacific Corp. in North Carolina (Region 4), was later withdrawn.

period over which the cost of Fund-lead actions will be spread. The assumptions we use in our model to estimate the cleanup cost of sites added to the NPL from FY 2000 through FY 2009 are based, in large measure, on the building blocks described in this chapter.

Chapter Four

Post-Construction Activities

Post-construction activities have not historically received the same level of attention as the remedial components of the Superfund program. In large measure, this is because the U.S. Environmental Protection Agency (EPA) has, understandably, placed a higher priority on implementing remedies than on the activities necessary after remedies are in place. In addition, the term "construction completion" is often misinterpreted to mean that site cleanup is *done*. In fact, a number of important activities are necessary to ensure that the physical remedies in place (1) achieve the cleanup goals envisioned for the site and (2) remain protective of public health and the environment. These activities typically include the long-term operation of groundwater pump-and-treat systems and the maintenance of landfill caps, the cost of which can be significant, especially at mega sites.

EPA faced persistent criticism in the first decade of the Superfund program for the seemingly slow rate of site cleanups, which was typically gauged by the number of sites deleted from the National Priorities List (NPL). The Agency developed the construction completion milestone in 1990[1] as an alternative measure of progress at Superfund sites. To many people, the number of construction-complete sites has replaced the number of site deletions as the yardstick for Superfund program accomplishments, and it has also become a central measurement tool for EPA.*

Post-construction activities begin at an operable unit once the engineering remedy has been completed, which may be before or after a site reaches construction completion. (EPA designates a site construction-complete only

---

*EPA uses the number of construction completions as a Government Performance and Results Act target/annual performance goal.

when the *last* remedial action at the site has been constructed.) Although an operable unit might have the engineering remedy in place, this does not necessarily mean that the remedial objectives for that operable unit have been achieved. In actuality, remediation may last decades, and post-construction activities may be required both until and after cleanup goals have been met.

For instance, at an operable unit where *in situ* bioremediation is selected to treat contaminated soil, the operable unit will reach the post-construction phase once the bioremediation treatment unit has been built. However, remedial action will not be completed until cleanup goals—as specified in the record of decision (ROD)—have been achieved and the final remedial action report is approved. This period may be 20 or 30 years, or even longer at some sites.

Post-construction activities include operation and maintenance efforts, implementation and monitoring of institutional controls (e.g., land or water use restrictions), groundwater monitoring, soil sampling, erosion control, leachate collection, lawn mowing, and fence repairs. Also included are five-year reviews, which are conducted by EPA to assess whether remedies are protecting human health and the environment; these reviews are sometimes required before a remedy is fully in place.

In this chapter we focus on two post-construction activities: long-term response action (LTRA), which is a specific type of operation and maintenance activity, and five-year reviews. Both activities were identified as important elements by the congressional report language requesting this study.[2] EPA is responsible for paying for and conducting (generally through a contractor or a cooperative agreement with a state) both activities. We also briefly discuss the limited information known about the cost of other post-construction activities. Lastly, we examine the use of institutional controls based on data we collected as part of our evaluation of five-year reviews. Specifically, we explore the extent to which institutional controls have been successfully implemented.

Long-Term Response Action

LTRA refers to a specific category of operation and maintenance activities, defined as the "Fund-financed operation of ground water and surface water *restoration* measures, including monitored natural attenuation, for the first ten years of operation" (emphasis added).[3] Not all groundwater and surface water remedies fall into the category of LTRA, as at many sites these remedies are designed for containment purposes only. Although LTRA usually refers to the operation and maintenance of pump-and-treat groundwater systems, it may also include monitored natural attenuation and *in situ* methods of cleanup, such as bioremediation, air sparging, and reactive walls.

Whether operation and maintenance activities are for groundwater restoration or for containment purposes is crucial in determining who pays for them. Although the operation and maintenance of Fund-lead remedial actions are the responsibility of states, LTRA is an exception. The National Contingency Plan (NCP) specifies,

> For Fund-financed remedial actions involving treatment or other measures to restore ground- or surface-water quality to a level that assures protection of human health and the environment, the operation of such treatment or other measures for a period of up to 10 years after the remedy becomes operational and functional will be considered part of the remedial action. Activities required to maintain the effectiveness of such treatment or measures following the 10-year period, or after remedial action is complete, whichever is earlier, shall be considered O&M [operation and maintenance].[4]

In other words, the financial and operational responsibility for the first 10 years of remedial and other measures to restore groundwater and surface water belongs to the federal government at those operable units where the remedial action was Fund-lead. As is the case with the cost-sharing arrangement for Fund-lead remedial actions, EPA pays 90% of the cost during the LTRA period; the state where the site is located picks up the balance. LTRA generally begins one year after a remedy has been constructed and is complete when cleanup goals are achieved or after 10 years, whichever comes first. If after 10 years cleanup goals have not been met, LTRA becomes the financial and operational responsibility of the state, consistent with the state's responsibility for operation and maintenance of all Fund-lead remedial actions.

Although LTRA is clearly defined in the NCP, there seems to be considerable confusion about what activities it encompasses. Moreover, the incidence and costs of LTRA have not been consistently captured in EPA's major management and accounting databases (the Comprehensive Environmental Response, Compensation, and Liability Information System, CERCLIS, and the Integrated Financial Management System, IFMS). In fact, we discovered that one EPA regional office had never heard of LTRA and as a result had for years been miscoding its response actions in CERCLIS.

LTRA Now and in the Future

At the end of FY 1999, there were 72 nonfederal NPL sites with LTRA under way.* This number will steadily increase in the coming years as more operable units reach the post-construction stage of the pipeline. The question that must be addressed to estimate future post-construction costs is how many Fund-lead remedial actions are likely to include the restoration of groundwater or surface water as part of the remedy—that is, how many operable units are likely to require LTRA in the future.

In an internal 1995 study of RODs signed from FY 1982 through FY 1995, EPA found that about 55% of remedies at nonfederal NPL sites included groundwater restoration.[5] For the RFF remedial action cost database,† we collected data on LTRA, in part to look at this issue. For each Fund-lead remedial action contract, we asked remedial project managers whether the remedy would be followed by LTRA. Of the 107 Fund-lead remedial action contracts awarded from FY 1995 through FY 1999 for which we had data, we found that 36% (38) included groundwater or surface water restoration as part of the remedy. Lacking more comprehensive data, we assume in our estimates of future costs (presented in Chapter 7) that 45% of Fund-lead remedial actions are likely to require LTRA. This estimate of the frequency of LTRA is about the midpoint between EPA's analysis and the data we gathered in the RFF remedial action cost database, and it is consistent with that used in prior studies of Superfund costs.[6]

The NCP limits LTRA to 10 years, but it is not uncommon for groundwater or surface water restoration to take decades before reaching cleanup goals. Thus, although LTRA is by definition finite, states will incur responsibility for financing and operating restoration measures at some sites for an extended period—conceivably forever if cleanup goals are not met. At one of the mega sites for which we gathered site-specific information, groundwater restoration (in this case, through monitored natural attenuation) was estimated to be needed for more than 200 years to attain the cleanup goals specified in the ROD. On the other hand, in some circumstances, restoration will be achieved in fewer than 10 years. Looking at the 35 LTRAs to be conducted at nonmega sites for which we had duration data as part of the RFF remedial action cost database, we found that the average planned duration for LTRA was approximately nine years.

---

*This includes only Fund-lead actions. PRP-lead groundwater restoration—known as PRP LR—was under way at 101 nonfederal NPL sites.

†See Appendix D for a detailed description of the RFF remedial action cost database.

Cost of LTRA

According to expenditure data in IFMS, EPA spent approximately $12.8 million on LTRA activities in FY 1999, or an average of $168,000 per LTRA. This is almost certainly an underestimate of the actual annual cost of LTRA. Prior studies by EPA and the U.S. General Accounting Office (GAO) estimated the average annual LTRA cost per site at $434,000 and $337,000, respectively.[7] Although these estimates come from data provided in RODs, they suggest that the average computed from IFMS is too low.*

We calculated an average annual LTRA cost for nonmega sites from data compiled in the RFF remedial action cost database. As part of our survey on remedial action costs, we asked remedial project managers to specify when a remedy called for either groundwater or surface water restoration and, when possible, to separate the cost of LTRA from the cost of the other parts of the remedial action. Information about these remedies provided a small (38) but recent sample of data on LTRA costs. Based on these data, the average annual cost for LTRA at a nonmega site is approximately $450,000.† The average annual cost for LTRA at mega sites is slightly more than $2 million, but only six sites are represented in the dataset.

We also analyzed data from a recent EPA report that evaluated the operation of groundwater treatment systems at 28 sites, 24 of which were NPL sites.[8] One criterion for inclusion in the Agency's study was that the groundwater treatment systems had to be at sites for which aquifer cleanup goals had been established, which makes the systems comparable to those used for LTRA. The average annual cost of operating these systems at nonmega sites and mega sites is $355,000 and $3.3 million, respectively, although the average for mega sites in the EPA study is based on only three sites.

Based on these analyses, our best estimate is that the average annual LTRA cost for an operable unit is about $400,000 at a nonmega site and $2 million at a mega site. These estimates do not include any future cost savings that might materialize from improvements in pump-and-treat groundwater remediation methods. After completion of groundwater optimization pilot projects in two EPA regions, the Agency and the U.S. Army Corps of Engineers recently concluded that annual LTRA costs could be reduced by an

---

*LTRA expenditures are most likely being captured in IFMS as RA expenditures, which would suggest that our RA unit costs are slightly inflated. Even if this is the case, however, the impact on our RA unit costs would be minimal.

†There did not appear to be much variation in LTRA costs among site types, although this may be due to the small number of examples for each site type.

average of 30%.[9] EPA has begun to conduct similar pilot projects nationwide, but the true extent of these potential cost savings remains an open question. In the coming years, as more operable units move into LTRA, the cost of LTRA should become clearer.

As noted earlier, if cleanup goals have not been achieved after 10 years, LTRA becomes the financial and operational responsibility of the relevant state as part of its operation and maintenance obligations. As of the end of FY 1999, LTRA had shifted to states in only a small number of cases. However, states are expected to bear responsibility for an increasing number of LTRAs in coming years. A common theme that emerged from our interviews with state officials was that although most have budgeted some money for operation and maintenance activities, the future expense of operating pump-and-treat systems for the restoration of groundwater will strain existing resources. Several state officials said they would try to persuade EPA to conduct additional cleanup (by reopening RODs) as a way to delay the shift of the financial burden of LTRA from the federal to the state government. In one state, the annual cost of LTRA at a single mega site is currently $4 million to $5 million, and the state officials we interviewed said they did not know where they would find the resources to pay for it when the time came to take it over. In recent years, states have lobbied Congress to shift most of the costs of operation and maintenance activities to EPA.[10] Clearly, one important emerging issue facing the Superfund program is the financial capacity of states to pay for the type of operation and maintenance often associated with LTRA.[11]

Five-Year Reviews

The Comprehensive Environmental Response, Compensation, and Liability Act of 1980 (CERCLA) instructs EPA to select permanent remedies wherever possible, preferably remedies that involve the treatment rather than the containment of hazardous substances. To provide a safeguard for those sites at which hazardous substances are not fully treated or removed, Congress included as part of the 1986 Superfund Amendments and Reauthorization Act (SARA) a requirement that every five years EPA verify that the selected remedy still protects human health and the environment. CERCLA states,

> If the President [EPA] selects a remedial action that results in any hazardous substances, pollutants, or contaminants remaining at the site,

the President [EPA] shall review such remedial action no less often than every 5 years after the initiation of such remedial action to assure that human health and the environment are being protected by the remedial action being implemented.[12]

Nonpermanent remedies (e.g., landfill caps, slurry walls) leave hazardous substances on-site for years—in many instances, in perpetuity. The five-year review requirement represents recognition that sites with nonpermanent remedies must be periodically inspected to ensure that people are not exposed to unsafe levels of contamination and that the surrounding environment is protected.

Five-year reviews, although relatively inexpensive to conduct, are an important post-construction activity at Superfund sites. Because these reviews provide information about whether EPA or responsible parties will have to revisit remedies, and what kinds of costs might consequently be incurred, the results are a window into potential future liabilities. Five-year reviews not only assess whether remedies are protecting public health and the environment but also allow EPA to evaluate remedies, correct problems and deficiencies, and adjust operation and maintenance efforts. This is especially important given the trend toward the selection of remedies where hazardous substances remain on-site above levels deemed safe for unrestricted use after completion of the remedy (referred to as *containment remedies*). As noted in a recent review of the Superfund program prepared for the National Academy of Public Administration, "[h]ow—and whether—those [containment] remedies will remain protective into the distant future is another question altogether. It is, perhaps, the major challenge facing the Superfund program in the next century."[13]

EPA has the final approval authority for all five-year reviews, although other government agencies (such as the U.S. Army Corps of Engineers and state environmental agencies) often perform the reviews through contracts or other agreements with EPA.[14] According to EPA's original five-year review guidance issued in 1991, the purpose of a five-year review is "to evaluate whether the response action remains protective of public health and the environment." The guidance indicates that there are two major goals of five-year reviews: first, to confirm that the remedy as spelled out in the ROD and/or remedial design remains effective at protecting human health and the environment and, second, to evaluate whether original cleanup levels remain protective of human health and the environment.[15] Each five-year review report is supposed to conclude with a section in which EPA includes a summary statement, referred to as the "statement of protectiveness," indicating whether remedies at a site are protective or, in other words, whether the two goals are being achieved.

To accomplish the first goal, EPA is to review the operation and maintenance of the site, conduct a site visit, and perform a limited analysis of site conditions. The focus of a five-year review depends on the original objective of the response action. If, for instance, the cleanup remedy is to provide protectiveness with a landfill cap and institutional controls, the review focuses on whether the cap remains effective and the institutional controls have been implemented. If, instead, the remedy calls for the long-term operation of a pump-and-treat groundwater system, the review focuses on the effectiveness of the technology and determines whether progress is being made toward the performance levels established in the ROD. Finally, all the reviews are to consider the performance and sufficiency of operation and maintenance activities.[16]

To establish whether the cleanup levels specified in the original ROD still provide adequate protection of human health and the environment (the second goal of five-year reviews), EPA is to analyze newly promulgated or modified requirements of federal and state environmental laws. Cleanup remedies are chosen at Superfund sites based, in part, on whether their implementation will achieve standards that are consistent with all applicable or relevant and appropriate requirements (ARARs). The purpose of this part of the review is to determine whether federal and/or state environmental standards have been newly established or modified to the extent that the original remedy no longer meets ARARs. For instance, if a maximum contaminant level for groundwater has been made more stringent by a state, the protectiveness of the remedy may be called into question.[17]

There are two types of five-year reviews: statutory and policy. Statutory reviews refer to the requirement set forth in section 121(c) of CERCLA and are further specified in the NCP.[18] As noted above, these reviews are required when, after a remedial action, hazardous substances, pollutants, or contaminants remain above levels that allow for unlimited use and unrestricted exposure. EPA also undertakes five-year reviews as a matter of policy at (1) sites where the remedial action will require five or more years to meet cleanup levels, even if the remedial action will result in the site's being open to unlimited use and unrestricted exposure; and (2) sites addressed prior to the 1986 SARA amendments, at which the remedy, upon attainment of the ROD cleanup levels, will not allow unlimited use and unrestricted exposure. EPA may also conduct five-year reviews at sites deleted from the NPL if the Agency deems it appropriate, as well as at non-NPL sites.[19]

EPA's 1991 five-year review guidance outlined three levels of five-year reviews. The appropriate level is determined by site-specific considerations, including the nature and status of the response actions, the proximity to populated areas and sensitive environmental areas, and the interval since the last review was conducted. A level 1 review is the standard version and is appropriate for most sites. A level 2 review includes recalculation of the risk presented by a site, and a level 3 review involves calculating a new risk assessment. EPA added a fourth level of review in its 1994 supplemental guidance on five-year reviews—level 1a—that is less comprehensive than the other types. The level 1a review is intended for sites where response is ongoing, whereas the other levels are intended for sites where response is complete (i.e., all remedies have been implemented).[20]

The trigger date for a five-year review is specified in EPA guidance and differs by review type. Statutory reviews are to be completed within five years of the initiation of the first remedial action at a site; policy reviews are to be initiated within five years of the date a site reaches construction completion.* Five-year reviews are performed at the site level and are to be conducted indefinitely at all sites at which hazardous substances are left on-site at levels that do not allow for unlimited use and unrestricted exposure following completion of the remedial action.

Five-year reviews have historically not received much attention, although this has begun to change.[21] GAO conducted a limited analysis of five-year reviews as part of a study on operation and maintenance.[22] The EPA Office of the Inspector General (OIG) has performed the two most comprehensive evaluations of the five-year review program. In its 1995 evaluation, OIG found a large backlog of five-year reviews, due mostly to the low emphasis given to them by Agency management. OIG stated, "Many officials we interviewed appeared to view the five-year reviews as a nuisance, and gave the impression that the reports had little or no value."[23] OIG conducted a second study four years later and found that although EPA had made some improvements in the five-year review program, the backlog of overdue reviews had almost tripled. OIG also concluded that the five-year review reports needed to be more informative and more effectively communicated to the public.[24]

---

*For reviews conducted early, the subsequent review is due within five years of the completion of the early review; if a review is conducted after the time it is due, the subsequent review is due within five years of the time it was originally required. U.S. EPA, *Supplemental Five-Year Review Guidance*, July 26, 1994, p. 2.

Since the 1999 OIG assessment, EPA has been drafting new guidance intended to improve the quality of five-year reviews, and the Agency has also committed to eliminate the backlog of overdue reviews by the end of FY 2002. Through FY 2000, EPA had completed a total of 618 five-year reviews at non-federal NPL sites and had made significant progress toward eliminating the backlog. In FY 2000 alone, the Agency completed 183 five-year reviews and reduced the backlog by half, to 69 overdue reviews.[25]

Cost of Five-Year Reviews

EPA does not have an official estimate of the cost of performing a five-year review. In the 1991 five-year review guidance, the Agency projected that a standard level 1 review would require about 160 to 170 hours of staff time, which was about $6,000, or approximately one month of full-time effort.* This estimate, however, does not include other costs that may be incurred during a review, such as those associated with site visits and sampling.

In its 1999 evaluation, OIG placed the cost of conducting a five-year review through an interagency agreement or a contract at $20,000 to $25,000.[26] We attempted to compute an average cost of five-year reviews using IFMS expenditure data. The code for five-year reviews, however, has been in IFMS only since 1996, and the number of five-year reviews represented in the system is considerably smaller than the number of reviews actually completed in the same period. As of the end of FY 1999, only 67 sites had expenditures charged to five-year reviews, and for more than 30% of these, the total is less than $1,000, indicating that the Fund dollars being used for five-year reviews are almost certainly not being captured correctly.† Lacking better data, we rely on OIG's estimate of $25,000 as the best current estimate of the cost of conducting a five-year review at a nonmega site.

Despite the coding problems in IFMS, our analysis of the expenditure data suggests that the cost of conducting a five-year review for a mega site is much greater than for a nonmega site, which is not surprising, since mega sites are generally larger and have more operable units. Although the number of mega

---

*Level 2 and level 3 reviews—which by definition are more rigorous—were projected to require 725 to 790 hours ($34,000) and 1,620 to 1,730 hours ($68,000) of staff time, respectively. U.S. EPA, *Structure and Components of Five-Year Reviews*, May 23, 1991, Attachment, p. 11.

†Five-year review expenditures are most likely being captured in IFMS as part of oversight or general response support.

sites for which data exist in IFMS is too small to calculate a reliable average, the cost typically surpassed $25,000 and for some sites is much higher (for example, $144,000 for the Denver Radium site and $62,000 for the Nyanza Chemical Waste Dump site). To account for this difference, we assume in our estimates that the average cost of conducting a five-year review at a mega site is $50,000, double that of a review at a nonmega site.

Compared with the cost of remedial pipeline actions and LTRA, the cost of a five-year review is small. However, the findings of a five-year review could imply significant future costs if remedies are either not protecting human health and the environment or deficient and in need of repair. For this reason, we performed an analysis of recently conducted five-year reviews; our findings are summarized in the next section.

Analysis of Five-Year Reviews

The main objective of our analysis of five-year reviews was to obtain a better sense of the potential long-term liabilities that might emerge at NPL sites. Although it is too soon to know the extent to which remedies will have to be revisited to ensure that cleanup objectives are met, let alone the expense of implementing any corrections or changes, analyzing five-year reviews helps shed some light on this issue.

Our analysis focused on the findings regarding cleanup remedies at NPL sites. In addition to evaluating whether the remedies selected are protecting human health and the environment, five-year reviews also provide valuable information about the effectiveness of remedies, changes in ARARs, and the adequacy of operation and maintenance activities. We examined five-year review reports for answers to four questions: (1) Was the remedy "protective of human health and the environment"? (2) Was the remedy complete? That is, did it include all components listed in the site's ROD(s)? (3) Was the remedy functioning properly? and (4) Were there significant changes in ARARs since the remedy was implemented that affect the protectiveness of the remedy? The results of our analysis are briefly discussed below.

We analyzed 151 five-year review reports for nonfederal NPL sites, representing most of the reviews completed in FY 1999 and through the first half of FY 2000. These 151 sites included all 11 RFF site types and a proportional number of mega sites,* and they represented each of the 10 EPA regions, although

---

*Of the 151 reports reviewed, 17 were for mega sites, which is approximately the same proportion of mega sites relative to all NPL sites.

they were not evenly distributed. More than 30% of the reviews were for sites in Region 5, for instance, whereas Region 6 completed only one five-year review in all of FY 1999. Of the 151 reviews, 51 were level 1a, 99 were level 1, and 1 was level 2. Much of our analysis focused on the 100 level 1 and level 2 reviews, since by definition these sites should have all of their remedial actions in place. The only information we considered in our analysis was that contained in the written report, which is the official public record of the five-year review.

For each of the 151 five-year reviews, we analyzed the statement of protectiveness, which is made at the end of each report. We classified these statements into six categories, which are presented in Table 4-1. The Agency determined in 99 of the 151 reviews we examined that the remedies at the sites were protective of human health and the environment. Of the 100 level 1 and level 2 reviews, conducted for sites at which all remedies are supposed to be in place, nearly 80% were found by EPA to be protective. As discussed in detail later, remedies were found not to be protective at 5% of these 100 sites. For the balance, EPA determined that either the remedy would be protective only after particular measures were taken or that it was unable to determine protectiveness. In some cases, the report failed to include a definitive statement.

We also examined the reports to determine whether all components of a site's remedial action(s)—as identified in the ROD(s)—were in place and functioning as designed. We found that in 75% (69) of the 92 level 1 and level

Table 4-1. Five-Year Review Statements of Protectiveness

	Reports		
	Levels 1 and 2	Level 1a	Total
Remedy remains protective	78	21	99
Remedy not protective	5	2	7
Remedy not yet fully in place but will be protective	1	24	25
Remedy will be protective if certain measures are taken	5	0	5
Unable to determine whether remedy is protective	5	1	6
No definitive statement	6	3	9
Total	**100**	**51**	**151**

Source: RFF analysis of 151 five-year review reports for nonfederal NPL sites completed in FY 1999 and FY 2000.

2 reviews for sites at which there were remedies,* all remedies at the site were fully implemented. For the other 25% (23) of the sites, the reports did not document full implementation. In two-thirds of these cases, the missing or incomplete parts of the remedies were institutional controls. In the remaining cases, the site was currently undergoing or needed further study or cleanup.

We found that 83% (76) of the level 1 and level 2 reviews for sites with remedies reported that the remedies were functioning as designed, but 17% (16) indicated some deficiency. These deficiencies included cases where remediation was taking longer than anticipated, contamination levels still exceeded standards, contamination was not being contained, or there were potential or actual structural problems with the remedy (beyond what would be considered repairable as part of typical maintenance).

Another key component of a five-year review is the examination of ARARs to determine whether there have been significant changes to federal or state laws and regulations that affect the protectiveness of the remedy selected in the ROD(s). Of the 100 level 1 and level 2 reviews, 72 included a review of ARARs. We found that there had been significant changes in ARARs at nine (13%) of these sites, although according to the reports these changes only brought the protectiveness of the remedy into question at two sites. In most cases, the remedy selected in the ROD(s) was determined to still meet the standards of relevant ARARs. The remaining 28 reviews did not include information in the report indicating that ARARs had been reviewed.

There does not seem to be a direct connection between whether EPA concluded that the remedy at a site was protective of human health and the environment and whether the five-year review report indicated that all remedies at the site were implemented and functioning as intended. In other words, EPA often is not following its own guidance as to what is meant by having protective remedies. At 12 of the 99 sites that EPA found to be protective, the five-year review report did not provide verification that all of the components of the selected remedies required by ROD(s) were in place. Interestingly, at all but 2 of these sites, the missing component of the remedy was institutional controls, a subject discussed later in this chapter. With respect to the relationship between protectiveness and the functioning of remedies, EPA concluded that remedies were protective at 7 other sites, even though information in the five-year review report suggested significant deficiencies with one or more of the remedies.

---

*Eight of the 100 level 1 and level 2 reviews were conducted for sites that only had "no further action" RODs.

The explanation for this discrepancy between the statement of protectiveness and full implementation and performance of remedies lies in the fact that the reports often have inadequately supported statements of protectiveness. This was one of the principal findings of OIG's 1999 study of five-year reviews[27] and is an issue that has been raised by others who have examined EPA's five-year review program.[28] Of the 99 reports that concluded remedies were protective, we found that almost half (48) of the statements were insufficiently substantiated or were questionable, based on evidence in the reports indicating that remedies were not fully in place, not functioning as intended, or not likely to achieve remedial objectives in a timely manner or, in some cases, at all. In 19 of these 48 reports, remedies had not yet been completed (all level 1a reviews), which suggests that it was premature to reach the conclusion that the remedies were protective.

Sometimes, even at sites where the five-year review report clearly indicated that a remedy was not working, EPA concluded that a site was protective of human health and the environment. For example, in the five-year review for the CTS Printex site in Mountain View, California, the report states,

> Computer modeling based on available data predicts that mass removal using extraction wells in the Printex Source area will no longer be effective by the year 2006. The model also predicts that 4.43 pounds of TCE [Trichloroethylene] will remain in the groundwater and that groundwater concentrations will reach asymptotic levels of approximately 54 ppb [parts per billion]. This indicates that the 5 ppb cleanup standard is unlikely to be met in the Printex area.[29]

The report subsequently recommends that CTS, the responsible party at the site, continue implementing the approved remedial actions and concludes with a statement certifying that the remedy selected for the site remains protective of human health and the environment, even though the language in the five-year review report clearly suggests that the remedy is unlikely to achieve cleanup goals. One would think that the findings would suggest a more comprehensive review of the remedy and consider whether a different remedy needs to be implemented. An additional issue regarding this particular five-year review is that it was performed more than three and a half years before EPA approved it.

Seven reviews concluded that the remedies at the sites were *not* protective of human health and the environment. Among the reasons remedies were found not to be protective were failure to implement institutional controls, improperly functioning remedies, changing characteristics of contaminants at the site, and discovery of new hazardous substances at the site. Although two of the reviews were level 1a, meaning that all remedies for the site had not yet been completed, in each it was found that remedies already initiated were not protective. One might expect to find these inadequate remedies at sites addressed early in the Superfund program, but this appears not to be the case. Of these sites, two had their first ROD signed in the 1980s, and five had their first ROD signed in the 1990s.

According to EPA guidance, when remedies are found not to be protective, the five-year review is to include recommendations to ensure that the remedy will become protective; set milestones toward achieving protectiveness, with clear timetables; indicate who has responsibility for carrying out the recommendations; and identify the agency responsible for oversight.[30] Although all seven of the reviews that concluded remedies were not protective included recommendations, milestones were clearly established in only three reviews; the party responsible for carrying out the recommendations was specified in five; and the agency in charge of oversight was named in only four. More importantly, it is not readily apparent how closely these recommendations have been followed, or even whether there is an internal process at EPA to ensure that the recommendations made in five-year review reports are systematically tracked and carried out.

There is not yet sufficient information to gauge the extent to which some remedies will be found inadequate in the future. Clearly, however, there are examples of sites where five-year reviews have uncovered problems with remedies or the need for further remediation, requiring significant additional investment from EPA or from responsible parties. At the Denver Radium site, for example, a five-year review completed in December 1999 for the Shattuck Chemical operable unit raised technical concerns about the ability of the monolith (solidified soils) cover design to meet the long-term performance criteria specified in the ROD.* Although the remedy was found "currently protective of human health and the environment," the long-term protection provided by the cover design could not be reasonably assured, according to the EPA review.[31] In addition to the technical concerns raised by the five-year review report, Colorado, the county and city of Denver, elected officials, and

---

*The report also identified deficiencies related to the lack of institutional control of the groundwater plume outside the site's boundary, the plume-monitoring plan, site characterizations and modeling, risk assessment, and other design issues, which raised additional technical concerns with the remedy, but these did not rise to the same level of significance as those for the monolith cover system.

members of the local community appealed to EPA to consider an alternative to the on-site remedy.[32] EPA responded in June 2000 by amending the ROD for the operable unit and changing the remedy from on-site containment to removal of the monolith and any additional contaminated soils to an off-site, licensed facility.[33] The Agency estimated that the removal would have a capital cost of about $20 million.[34]

The findings of the five-year review for the Tar Creek site in northeastern Oklahoma provide another example of the potential cost implications of remedies that have to be revisited. In this case, the review identified the need to conduct additional cleanup to address lead-contaminated soil on residential properties. Although the original remedy had been completed, the target blood lead level in children was lowered by the Centers for Disease Control and Prevention, and health studies reported a significant number of children living in neighborhoods near the site with blood lead levels above the new target level.[35] By the end of FY 1999, EPA had spent an additional $25 million to clean up the lead-contaminated soil through a combination of removal and remedial actions,[36] and the Agency will need to spend tens of millions more to complete the cleanup effort.

These examples illustrate the role that five-year reviews can have in uncovering problems with remedies and revealing potential liabilities. There is nothing to suggest that EPA will often be confronted with the level of costs associated with the Denver Radium and Tar Creek sites, both of which are Fund-lead mega sites. However, it is inevitable that remedies at some sites will be found to be not functioning as designed or not adequate (in the short term or the long term) to fully protect human health and the environment, and will thus require the investment of additional funds in the future.

Other Post-Construction Activities

In addition to LTRA and five-year reviews, a number of other important post-construction activities must be conducted after a remedy is in place. These activities include review of reports (e.g., groundwater monitoring), implementation and monitoring of institutional controls, on-site inspections, oversight of remedies and operation and maintenance, performance of remedy optimization studies, and administrative activities associated with site deletion and reuse efforts.[37]

EPA has limited historical information about costs incurred at sites after they are construction-complete but did provide some estimates based on its analysis of IFMS expenditure data and discussions with Region 5, which has an organizational unit focused on post-construction. Region 5 estimated that per year per site, post-construction activities cost about $32,000, which covers staff time, site visits, and laboratory expenses. There also can be one-time expenses for other studies (e.g., optimization, technical impracticability, modeling), which have costs on a par with five-year reviews—$20,000 to $25,000. A final set of extramural costs includes those for repair or replacement of Fund-lead groundwater pump-and-treat system components used during LTRA before their transfer to the state.[38]

Implementation of Institutional Controls

Our analysis of five-year reviews provided an opportunity to collect data on the types of institutional controls used at NPL sites and to examine the extent to which they have been successfully implemented. Institutional controls at Superfund sites are typically restrictions placed on land or water use. The general purpose of an institutional control is to separate the public from levels of contamination that potentially pose an unacceptable health risk. Institutional controls are used when it is not cost-effective or technically feasible to reduce the volume of contamination to levels that would allow unrestricted use of the site.[39] There are two main categories of institutional controls: governmental and proprietary. Governmental controls include a wide range of land use controls that local governments can impose, such as zoning restrictions and building permits. Proprietary controls are based on the rights associated with private ownership and include deed restrictions and easements. Most often, institutional controls are one component—rather than the only component—of a site remedy.

Our examination of the five-year review reports found that of the 151 sites, 58% (87) included at least one institutional control as part of the response at the site. If the definition of institutional controls is expanded to include passive controls, such as fencing, security guards, signage, and other administrative measures, almost 80% (119) of the 151 sites relied in part on an institutional control. If these data are representative of all sites on the NPL, it is apparent that the effectiveness of response depends on the successful implementation of institutional controls.

Of the 100 sites for which EPA conducted a level 1 or level 2 review—that is, at sites where all remedies should be in place—at least one institutional control (not including passive controls) was part of the remedy at 61. According to the reports, the 61 sites had a total of 81 institutional controls, or an average of 1.3 per site. Table 4-2 summarizes the types of institutional controls at these 61 sites, and indicates the implementation status of each, based on information provided in the five-year review reports. The single most common type of institutional control was a deed restriction, which was part of the response at 35 sites. Other frequently used controls were notices and advisories and zoning restrictions. Our analysis of the reports indicated that 75% (61) of the 81 institutional controls at these 61 sites were fully implemented. Of the remaining 20 institutional controls, 2 were partially implemented, 15 were not implemented, and 3 had an unknown status (their implementation status was not mentioned in the report). At the majority of these sites, at least one institutional control was specifically identified in the ROD(s). At the rest of the sites, it was either unclear from the information provided in the five-year review report whether the institutional control was included in the ROD or it was noted in the report that it was part of some other action or decision-document at the site.

Interestingly, at the 99 sites for which EPA determined in the five-year review that remedies were protective of human health and the environment, we found that 28% (13 of 46) of the institutional controls that were explicitly called for in the ROD(s) for these sites were either not fully implemented or had an unknown status. In other words, when selecting remedies for these sites, EPA determined that institutional controls were necessary components. Yet, when reviewing remedies to determine whether they were protective of human health and the environment, the agency often seemed to ignore the fact that institutional controls were not implemented, contradicting its own guidance on what it means to have a protective remedy. Failure to implement institutional controls means that the remedy is not complete and raises the question whether the remedy is sufficient to protect human health and the environment.

Conclusions

Post-construction activities are an important part of the Superfund cleanup program. Although EPA has historically placed less emphasis on these activities than on those associated with remedial pipeline actions, this has begun to change as more operable units are reaching the post-construction phase of the pipeline and more sites are designated construction-complete. The cost to EPA of implementing post-construction activities are relatively small, with the notable exception of LTRA at mega sites, which can be millions of dollars

Table 4-2. Status of Institutional Controls at 61 NPL Sites

Type of institutional control	Number	Percentage	Status			
			Fully implemented	Partially implemented	Not implemented	Unknown
Government						
Notice and advisory	7	9%	6	0	1	0
Permit	5	6%	5	0	0	0
Site acquisition	1	1%	1	0	0	0
Zoning restriction	14	17%	13	0	1	0
Not specified	9	11%	8	0	1	0
Proprietary						
Deed restriction	35	43%	21	2	10	2
Easement	1	1%	1	0	0	0
Other	3	4%	3	0	0	0
Not specified	6	7%	3	0	2	1
Total	**81**	**100%**	**61**	**2**	**15**	**3**

Note: Totals may not add exactly because of rounding.

Source: RFF analysis of 100 level 1 and level 2 five-year review reports for nonfederal NPL sites completed in FY 1999 and FY 2000; 61 sites had at least one institutional control.

annually. The ability of states to pay such costs, as an increasing number of LTRAs become their responsibility as part of their operation and maintenance obligations, is an issue that merits further examination.

Our evaluation of five-year reviews indicates that EPA needs to improve the quality of these reports as well as begin to track whether recommendations made in the reviews are being implemented. Moreover, it is clear that the agency, in violation of its own guidance, sometimes concludes that remedies are protective when remedies are incomplete, not functioning as designed, or unlikely to meet remedial objectives in the time frame anticipated at the time the remedies were selected for the site.

Our analysis of five-year reviews further suggests the importance of evaluating these documents, as they provide a harbinger of the costs that may need to be incurred in the future to ensure that remedies remain protective. Little is known about the future liabilities that may result from revisiting remedies that are not functioning as intended or the costs that will be incurred to fully implement and monitor institutional controls. In the coming decade, EPA and Congress will need to pay greater attention to post-construction activities as they assume a more central role in the Superfund program.

Chapter Five

The Size and Composition of the Future National Priorities List

In this chapter we estimate the number and types of sites likely to be added to the National Priorities List (NPL) in the next decade. We approach this task with some trepidation. Previous attempts to estimate the future size of the NPL by the Congressional Budget Office (CBO), the now-defunct congressional Office of Technology Assessment, and the U.S. Environmental Protection Agency (EPA) have proven far off the mark.[1] Instead of thousands of new sites being added to the NPL, as these studies predicted, only 140 sites were listed from FY 1995 through FY 2000. Our forecast of an upswing in the number of sites added to the NPL may prove equally unfounded.

The number of sites added to the NPL in the coming years is a key determinant of the program's future cost. However, the *number* of sites added to the NPL may be less important for estimating costs to the federal government than the kinds of sites, the number of mega sites among them, and the percentage of cleanup actions that will be Fund-lead. Our estimates are therefore based on three major components or cost drivers: the number of sites likely to be placed on the NPL through FY 2009, the type of contaminated sites that will be included on the NPL, and the extent to which the federal government will have to foot the bill.

To develop our estimates of the number and type of sites added to the NPL, we first examined the program's past in order to identify trends that might continue in the next two to five years. Relying solely on extrapolations from the past, however, may fail to account for emerging political and regulatory practices that will be felt more fully in the future, such as the impact of Superfund reforms on listing rates or shifts in state attitudes toward using the NPL to address contaminated sites. As one state official put it, "[t]he NPL is no longer reserved for the 'worst of the worst' sites, rather the NPL has shifted to a venue for remediating sites which require federal resources. The criteria for listing sites on the NPL may quickly shift from one of risk-based determinations to one based on resource needs."[2] Basing estimates on historical trends alone would also fail to take into account broader economic changes. For example, will an economic downturn in the next five years affect the viability of certain industrial sectors and facilities, making them, in effect, Superfund sites in waiting? If potentially responsible parties (PRPs) go out of business, will states have the financial resources to clean up these sites under state superfund programs, or will they end up on the NPL as Fund-lead sites? Alternatively, with a buoyant economy, will more state legislatures pass bond issues, as Michigan did in 1999, to fund state hazardous waste programs and to address orphan sites?

To understand more fully how policy choices and broader political and economic factors are likely to influence future NPL listings, we conducted detailed interviews with all 10 EPA regional Superfund division directors and 9 state hazardous waste officials,* and asked the following questions:

- What factors influence site listing?
- From the perspective of front-line EPA regional and state officials, how many and what types of sites will be listed on the NPL in the next two to five years?
- What will be the major cost drivers in the future?
- Are future NPL sites more or less likely to be mega sites—that is, sites with total removal and remedial action costs of $50 million or more—and are these sites likely to be Fund-lead or PRP-lead?
- Are future NPL nonmega sites more or less likely to have Fund-lead cleanup actions?

We begin this chapter with an overview of how sites are identified and evaluated for NPL eligibility, as well as a discussion of policy choices surrounding listing sites on the NPL. We then briefly review past efforts to estimate the size of the NPL universe and discuss why forecasting the future size and composition of the NPL is highly uncertain. In the final section of this chapter, we summarize the findings of our research, analyses, and interviews

---

*Under the Paperwork Reduction Act of 1995 (44 U.S.C. 3501–3520), RFF could conduct a maximum of nine interviews. A more extensive survey of state programs would have required approval from the Office of Management and Budget through a waiver provision; given the tight deadlines of the project, this alternative was not pursued.

and present three scenarios for the number of sites likely to be added to the NPL from FY 2000 through FY 2009.

The NPL Listing Process

The first step in the NPL listing process is the identification of sites that may pose risks to the public or the environment. Potentially contaminated sites are brought to the attention of EPA through a variety of channels: referrals from other federal regulatory programs, notification by state regulators, and petitions from local residents. In some cases, regional EPA or state staff identify sources of groundwater contamination or possible contamination from a particular industrial sector, such as abandoned coal gasification plants. Once such sites have been identified, EPA has the authority to take short-term immediate action under its removal program.

Figure 5-1 depicts the stages of the listing process. As the diagram suggests, not all contaminated sites identified are entered into the Comprehensive Environmental Response, Compensation, and Liability Information System (CERCLIS), EPA's inventory of potentially contaminated sites, which contains information on site assessment and remediation from 1983 to the present. Both EPA and the states can take the lead on contaminated sites in CERCLIS, but state hazardous waste programs also address sites that are not in CERCLIS through their state superfund or voluntary cleanup programs.

Most potentially contaminated sites are investigated by staff in state environmental agencies, not by the federal government. According to a 1998 report by the Association of State and Territorial Solid Waste Management Officials (ASTSWMO), state hazardous waste agencies addressed more than 27,000 sites from 1993 through 1997.[3] It is important to note that in the ASTSWMO report, the term "addressed" does *not* mean completed cleanups; it refers to a range of possible state responses including "no further action" designations, removal actions, site investigation, as well as site remediation. The vast majority of sites addressed by state superfund programs are not added to CERCLIS, although a proportion of these sites—data are not available to establish what percentage—are ultimately referred to EPA for evaluation. The states typically refer sites to EPA for action if PRPs are recalcitrant or not financially viable, if cleanup costs are anticipated to exceed state financial resources, or if the remediation may require relocation of residents.

Once staff in EPA regional offices are informed about a site, the pre-CERCLIS screening process begins. Implemented in 1998, this process is used

Figure 5-1. The NPL Listing Process

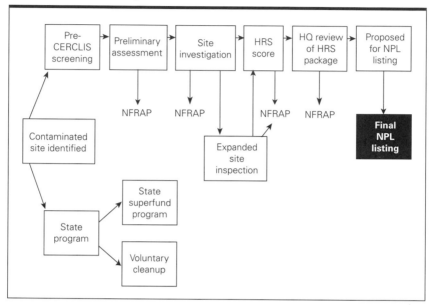

Note: NFRAP = no further remedial action planned; HRS = hazard ranking system; and HQ = headquarters (EPA)

to determine whether further site investigation is required under the Comprehensive Environmental Response, Compensation, and Liability Act of 1980 (CERCLA) and to minimize the number of sites unnecessarily entered into the CERCLIS inventory. According to the screening criteria, sites should *not* be entered into CERCLIS if:

- the site is currently in CERCLIS or has been removed from CERCLIS and no new data warrant CERCLIS entry;
- the site and some contaminants are subject to limitations in CERCLA, where the release is of a naturally occurring substance in its unaltered form, or in the case of releases in drinking water supplies due to deterioration of the system through ordinary use;
- a state or tribal remediation program is involved in the response, including the state superfund program, a voluntary cleanup program, and state or local brownfields programs;
- the hazardous substance release at the site is regulated under a statutory exclusion (e.g., petroleum, uranium mill tailings) or can be addressed

under another federal statute (e.g., corrective action under the Resource Conservation and Recovery Act, or RCRA);

- site data are insufficient to justify CERCLIS entry (further data collection is usually called for in these cases); or

- sufficient documentation exists that demonstrates there is no potential for a release that could cause adverse environmental or human health impacts.[4]

Once a site is entered into CERCLIS, EPA regional staff, state staff, or a contractor conducts a preliminary assessment (PA) of the site. A PA involves reviewing existing reports and documentation about the site, analyzing geological and hydrological data, and identifying populations and sensitive environments likely to be affected by a release of hazardous substances. At this stage, the goals of the PA are to establish whether a removal action(s) is necessary and whether the site poses potential risks to public health. If the PA determines that the site does not present a potential risk or contains contaminants not addressed under Superfund (e.g., petroleum), the site is typically eliminated from further consideration in terms of federal action,* or in the vocabulary of the program, NFRAPed—no further remedial action planned.

For those sites requiring further investigation, the next step in the process is the site inspection (SI). Although the PA is typically an off-site review, the SI involves a hands-on inspection in which soil and water samples are collected to better characterize site contamination. The SI is often conducted in two stages. In the initial stage, samples are analyzed to provide a more precise picture of which contaminants are present and to determine the potential for migration of releases of hazardous substances. If additional data are required, EPA will conduct an expanded site investigation (ESI), which can include complex background sampling and the installation of monitoring wells. If the SI or ESI determines that releases from the site pose no threat to human health and the environment, the site is NFRAPed.

The early stages of the screening process are relatively inexpensive. According to EPA, initial site screening costs about $4,000 per site. The next two phases of the screening process, the PA and SI, cost on average approximately $8,000 and $37,000, respectively. The ESI is the most expensive component of the screening process, with an average cost of about $100,000 per site. [5]

In the next step in the NPL listing process, data collected and analyzed as part of the PA and SI are used to develop a hazard ranking system (HRS) score. The HRS is EPA's screening tool to determine whether a site should be

---

*NFRAPed petroleum sites posing risks to human health and the environment are typically addressed by state programs.

considered for listing on the NPL. HRS scoring is the primary, but not the only, mechanism by which EPA places sites on the NPL.* It takes into account the likelihood that a site has released, or has the potential to release, hazardous substances into the environment; the toxicity and quantity of the hazardous substances at the site; and the proximity of people and sensitive environments to the release. The HRS scores four pathways of potential human exposure to contamination—groundwater, surface water, soil, and air—and then combines the individual pathway scores into an aggregate site score. Sites with an HRS score below 28.5 are NFRAPed. But if the preliminary HRS score equals or exceeds 28.5, the site is considered for inclusion on the NPL and goes through additional review and screening before it is actually proposed for listing.†

Many sites have become stalled at this point in the CERCLIS pipeline. As of October 2000, some 2,500 sites in CERCLIS had received a preliminary HRS score of 28.5 or above and were classified as "NPL eligible."[6] Only a small subset of those sites with a preliminary HRS score of 28.5 move on to the next step, where a formal listing package is prepared and the site is proposed for inclusion on the NPL. In this process, EPA regional staff submit an HRS package to EPA headquarters for review. After evaluating the data, EPA headquarters, in collaboration with the region and with the concurrence of the governor of the state in which the site is located, may propose the site for listing.‡ EPA then publishes the list of sites in the Federal Register as a proposed rulemaking, at which time the site listing decisions are generally subject to a 60-day public comment period. At the end of this period, after public comments have been considered and negotiations with

---

*There are two other mechanisms in addition to the hazard ranking system (HRS) that can be used to place sites on the NPL. Under CERCLA 105(a)(8)(B), each state may add a site to the NPL by designating it as its top priority, regardless of the HRS score. Forty sites have been added to the NPL under this provision. Also, sites can be placed on the NPL if the Agency for Toxic Substances and Disease Registry issues a health advisory that recommends people avoid contact with a release, EPA determines the release poses a significant threat to public health, and it is more cost effective for EPA to use its remedial authority than its removal authority to respond to the release. Thirteen sites have been listed on the NPL through this mechanism.

†An HRS score of 28.5 is the threshold for listing a site on the NPL. However, sites scoring higher on the HRS are not given priority in the NPL cleanup queue.

‡This provision is in the Omnibus Consolidated Rescissions and Appropriations Act of 1996, Public Law 104-134, enacted May 2, 1996.

other interested parties have been conducted, EPA compiles a final rule, which is signed by the Assistant Administrator of the Agency's Office of Solid Waste and Emergency Response (OSWER) and published in the Federal Register.* It is only at this point that a site is considered a "final" NPL site.

We have described the listing process as a rather straightforward series of events, but listing a site on the NPL is ultimately a policy choice made by the Assistant Administrator of OSWER, typically after discussions with EPA regional officials, state cleanup program staff, potentially responsible parties, and state and locally elected officials. The Agency's decision to list a site on the NPL is strategic, taking into account budgetary concerns, the political costs or benefits of listing, the capacity of the state to undertake cleanup, and the likely response of a PRP to the threat of listing, among other factors. Further complicating the matter, EPA Superfund managers and state officials can pursue alternatives to NPL listing, such as cleaning up sites under state programs or addressing a site as an NPL equivalent. The outcomes of these multiparty negotiations are uncertain. For instance, decisions to list a particular site may be intricately linked to negotiations about whether to list other contaminated sites in the same state. And, although the goal of NPL listing and Superfund cleanups is to reduce risks to human health and the environment from the release of hazardous substances, a number of interwoven issues make it difficult to predict whether a site will be listed on the NPL, addressed by a state program, or remain in limbo—a theme we explore more fully in the sections that follow.

Past Estimates

From the outset it should be emphasized that forecasting the future size of the NPL is a difficult and, some would say, foolhardy task. Evidence from Superfund's own history suggests that estimates of the future size of the NPL and future costs, particularly over a 10-year period, are highly uncertain and can be strongly influenced by available funding as well as expectations about future funding. In 1980, when CERCLA was enacted, Congress asked EPA to

---

*Not all sites proposed to the NPL become final. As noted in Chapter 3, some sites proposed for listing are addressed as NPL equivalent sites, under state superfund or other programs, or under other federal authorities.

create a list of "national priorities" and requested an initial list of at least 400 sites.[7] In response, EPA developed the HRS, the Agency's primary screening mechanism for placing sites on the NPL. The Agency selected an HRS cutoff score of 28.5 because it led to an initial NPL of 418 sites.[8] During the program's early years, according to an Office of Technology Assessment report, concerns about the long-term availability of Trust Fund monies led to an attempt to hold the size of the NPL to the original 418 sites, and to the strategy "there will be no Son of Superfund."[9] Superfund has had a number of progeny, however. Since the inception of the program, about 1,300 nonfederal sites have been proposed to the NPL, and in the past two decades some 43,000 sites have been added to CERCLIS. Although nearly three-quarters of these sites have been "archived"—meaning EPA no longer considers them candidates for NPL listing—there remains a backlog of some 6,400 sites in the CERCLIS pipeline, of which an unknown proportion will eventually be listed on the NPL. Table 5-1 shows the disposition of these sites in CERCLIS.

As of October 2000, 3,826 sites were in the early stages of the CERCLIS pipeline, with preliminary assessments needed or ongoing at 1,690 sites, site investigations needed or ongoing at 1,841 sites, and expanded site investigations needed or ongoing at 295 sites. At the latter stage of the pipeline, more than 2,500 sites had been assessed and were characterized as NPL eligible. Many of these eligible sites were added to CERCLIS over a decade ago and have been awaiting an NPL decision for several years.[10] EPA has yet to list these sites for a variety of reasons: The Agency considers the state to have the lead for site cleanup, EPA regional staff and state superfund officials have not determined the lead for cleanups, or data on current risks are needed and officials have to reassess data collected more than a decade ago.[11]

Clearly, the extent of the problem of contaminated sites was not easy to discern when CERCLA was enacted 20 years ago, but more recent estimates of the future size of the NPL are also marked by a high degree of uncertainty and vary considerably. Analyses by EPA and CBO in 1994 estimated that between 1,500 and 7,800 sites could be added to the NPL in the next 20 years.[12] These estimates, in effect, suggested a doubling of the size of the current NPL at the lower end of the range and a sixfold increase at the upper end, indicating that after 20 years of Superfund, much of the work remained to be done.

A more recent study by the U.S. General Accounting Office (GAO) and testimony by EPA officials have raised the possibility that the NPL universe may be much smaller than the earlier studies suggested. Taking a somewhat different tack than the CBO and EPA analyses, the 1998 GAO report, *Hazardous*

Table 5-1. Status of Potential NPL Sites

Stage in CERCLIS pipeline	Number of sites
Preliminary assessment ongoing or needed	1,690
Site inspection ongoing or needed	1,841
Expanded site inspection ongoing or needed	295
Assessment completed, NPL eligible	2,572
Total	**6,398**

Source: CERCLIS data provided to RFF by EPA, October 2000.

Waste: Unaddressed Risks at Many Potential Superfund Sites, tried to determine how many of the 3,036 sites in CERCLIS then classified as "awaiting a National Priorities List decision"—that is, sites potentially eligible for NPL listing with a preliminary HRS score of 28.5 or above—would ultimately be added to the NPL.* State and EPA officials identified 232 sites, or 8% of the total, as likely candidates for listing but could agree on only 26 sites, or less than 1% of the total number of NPL-eligible sites. This finding has been interpreted in a number of ways: that site-specific data in EPA's hazardous substance site inventory are inaccurate and outdated; that an HRS score of 28.5 or above is a poor predictor of which sites will ultimately be listed; and most importantly, that this apparently limited number of future NPL sites suggests that a ramping down of the Superfund program is occurring.[13] Less publicized, however, is perhaps the most significant finding of the GAO report: after having deferred decisions on a large number of sites in the CERCLIS pipeline, state and federal hazardous waste cleanup officials had yet to determine who would do the cleanup and under what program future cleanups will occur at more than half of the potentially eligible sites in the survey.[14] This underscores the difficulty of predicting what sites will actually be listed on the NPL. Moreover, the GAO report did not evaluate all the sites in EPA's inventory. Nearly 3,500 sites in the earlier stages of site assessment in the CERCLIS pipeline were not included in the study, nor did GAO attempt to forecast how many sites would end up on the NPL from future site discoveries.[15]

---

*Since the publication of the GAO report in 1998, state superfund programs and EPA regions have conducted additional site assessment work at many of the 3,036 sites GAO identified as NPL-eligible. Roughly 450 of these sites had been archived or NFRAPed, leaving 2,572 NPL-eligible sites in the CERCLIS inventory as of October 2000.

A further look into the CERCLIS backlog reveals that sites may get stuck in the pipeline because they are technically complex and costly, involve negotiations with recalcitrant PRPs, or are politically contentious. According to state and EPA officials interviewed for this study, sites may languish in CERCLIS while states pursue other cleanup options; in many (but not all cases) NPL listing comes only when these other options are exhausted. For example, the Landia Chemical Company site, a former pesticide manufacturing facility in Lakeland, Florida, listed on the NPL in May 2000, was first discovered and added to CERCLIS in 1983, when reports of pesticides in a drainage ditch near the site led the Florida Department of Environmental Protection to investigate. Because of enforcement problems and ultimately nonviable PRPs, the state discussed the situation with EPA in 1997 and requested that it be listed on the NPL. Cleanup costs will be paid from the Trust Fund and are expected to run between $15 million and $25 million.[16] This is just one example, but recent trends appear to indicate that sites in the CERCLIS pipeline for a decade or more can be a significant source of future NPL sites. Nearly one-quarter of the 112 sites listed on the NPL since 1998 were added to the CERCLIS inventory in the 1980s.[17] Clearly, some sites "age" and are added to the NPL years after they are first identified as sites of concern by state and federal officials.

Any forecast of the size and composition of the NPL in the next decade needs to consider what percentage of sites now being assessed in CERCLIS will be placed on the NPL. For this known universe of sites, as we have seen, estimates vary considerably among federal and state program managers. It is even more difficult to predict how many sites will be discovered and added to EPA's inventory in the next 10 years. The sources of this uncertainty are discussed below.

Sources of Uncertainty

Forecasting the future size and composition of the NPL is marked by three major sources of uncertainty:

- inadequate data identifying contaminated sites, due in part to the program's history of passive site discovery;
- difficulty in predicting the number of Fund-lead mega sites that will be added to the NPL—and the likelihood that some of these will be contaminated sediment or mining sites, which are among the most expensive to clean up; and

- the capacity of state hazardous waste programs to clean up sites that warrant NPL listing but are orphan sites (i.e., where the responsible parties are bankrupt or cannot be found) or have recalcitrant PRPs.

Unknown Size of the NPL Universe

After 20 years of Superfund, one might expect to have a clearer picture of the number of contaminated sites in the nation that warrant listing on the NPL, or at the very least a comprehensive list of potential Superfund sites. The EPA inventory of potentially contaminated sites, CERCLIS, in part serves this function, but states also maintain their own lists apart from CERCLIS, and the 9,500* active sites in CERCLIS are thought to be only a small portion of what is largely an unknown universe of contaminated sites needing attention. According to a survey of state hazardous waste officials conducted in 1998 by the Environmental Law Institute, states identified 69,000 "known and suspected sites."[18] GAO and others have estimated the number of contaminated sites in the country to range from 150,000 to 500,000, although only a small percentage of these sites are likely to warrant placement on the NPL.[19]

Information pertaining to contaminated sites is incomplete and inconclusive for a number of reasons. In part, it is not in the interests of property owners to disclose information about site contamination to regulators; such information could reduce the value of their property and make them liable for cleanup costs. Nor is it in the interest of most federal or state program managers to help create a comprehensive and public list of contaminated sites if they do not have adequate resources to address them.

The data deficit can also be explained in a historical context. CERCLA includes two provisions that relate to the identification of sites contaminated with hazardous substances: Section 103(c) sets forth requirements for notifying the federal government when sites contaminated with hazardous substances are identified, and section 105(a)(1) of CERCLA provides that the National Contingency Plan (the major regulation guiding the Superfund program) shall include "methods for discovering and investigating facilities at which hazardous substances have been disposed of or otherwise come to be located."[20] Other than that, however, Congress did not specify how a site dis-

---

*The 9,500 "active" sites in CERCLIS include not only the 3,826 sites needing assessment and the 2,572 sites characterized as NPL-eligible, but also 3,056 archive candidates.

covery program should work, and there is no active, coordinated site discovery program in the Superfund program. As noted earlier, contaminated sites are reported to EPA through a variety of channels from federal and state agencies and citizen complaints. In addition, some contaminated sites are identified through property transfer laws in states that require sellers of certain properties to inform state agencies about the environmental conditions of the site as a requirement of the sale.[21]

A more proactive approach to site discovery, however, would be resource intensive and require the allocation of more resources to this aspect of the Superfund program. In recent years the focus of the federal Superfund program has been achieving construction completion at sites already listed on the NPL, rather than finding new sites.[22] Similarly, a recent ASTSWMO survey found that with an increased workload, many state programs had made a policy decision not to spend state money on active site discovery.[23]

Forecasting the number of future NPL sites is, in essence, a problem of predicting demand, not supply. An aggressive, comprehensive, national site discovery program, utilizing historical aerial photographs and data from remote sensing and tied to a web-based geographic information system, would in all likelihood identify more contaminated sites. Combining federal and state inventories of contaminated sites and checking the data for inconsistencies, erroneous entries, and redundancies would help unify what has been a fragmented approach to site discovery. Yet without significant increases in resources and assurance of continued funding to address a potential influx of new sites into the program, Superfund managers and their state counterparts have little incentive and perhaps inadequate capacity to search for new sites, create a comprehensive national inventory, and determine through preliminary assessments and investigations which sites are appropriate for NPL listing.

In FY 1999, $57.8 million, or just 4%, of total Superfund expenditures went to site assessment activities. Of this total, 69%, or $39.8 million (mostly extramural dollars), was spent to conduct site assessments and inspections and, where appropriate, prepare the packages that nominate sites for listing.* Another $3.1 million was spent in EPA regional offices for management and support of site assessment activities, and $14.9 million was spent at EPA headquarters for the staff and contractors who develop national site assessment program policies and manage the national site assessment effort. Funding for

_____

*EPA estimates that it costs an average of $31,000 to prepare a listing package.

site assessments fell from $42.8 million in FY 1992 to $39.8 million in FY 1999, a 7% reduction. In constant 1999 dollars, this represents an 18% reduction.[24]

In short, the size of the NPL is not fixed but conditional, and the number of sites identified as potential NPL sites depends on policy choices EPA Superfund managers make, the level of site assessment funding, the strategic behavior of potentially responsible parties, the quality and robustness of state cleanup programs, and the extent to which EPA and states invest in site discovery. And even if more funds were devoted to site discovery and assessment, the question remains: What will be the role of the NPL in the national effort to address contaminated sites, and what sites should be listed? As noted earlier, for states with strong programs the NPL is no longer reserved for addressing only the "worst of the worst" sites, but rather, it is a means by which states can clean up sites that require federal resources. And for small states with less robust programs and limited financial capacity, the NPL may be the only way to finance the cleanup of sites that pose significant risks.

Unknown Number of Mega Sites

The number of sites added to the NPL is only one factor that will influence the amount of federal dollars required to clean up contaminated sites in the course of the next decade and beyond. As noted earlier, other important cost drivers are the type of sites placed on the NPL and whether cleanup actions at these sites will be Fund-lead or PRP-lead. For example, based on RFF's estimates of the average cost of Fund-lead remedial pipeline actions, the average cleanup cost for a mega site is roughly $140 million, more than an order of magnitude greater than the average cleanup cost of $12 million for a nonmega site.* The crucial question, then, is how much of the cleanup at future mega sites is likely to be Fund-lead. Undoubtedly, Fund-lead remedial actions at mega sites will be the single most expensive component in any estimate of future costs. And yet, as discussed below, no element of our cost estimates is more difficult to predict than the number of future Fund-lead mega sites. Most of the mega sites on the NPL were listed in the early years of the program. From FY 1983 through FY 1986, 12% of all sites (83 of 712) listed on the NPL were mega sites. Since then the number of mega sites has declined. From FY 1987 through FY 1999, only 5% of all sites (29 of 533) listed on the

---

*Estimates include only extramural costs for RI/FSs, RDs, and RAs and assume 3.8 operable units for a mega site and 1.6 operable units for a nonmega site.

NPL were mega sites. This decline may support what a CBO study called the "barrel-scraping effect."[25] The term suggests that EPA identified the worst and most costly sites in the earlier years of the program, and that less contaminated and less expensive sites, not mega sites, would constitute the bulk of the work for the program in the future.

Although the concept seems intuitively right—one would imagine that after 20 years of Superfund the worst sites would be addressed and there would be no new mega sites, or very few—it is difficult to determine from available data whether, in the words of one state official, all the "sleeping giants" have been addressed. More than one state and EPA official mentioned recent mega sites that, from their perspective, came out of nowhere—sites they referred to as "pop-ups." For example, in January 1999 EPA listed the Federal Creosote site in New Jersey on the NPL. This 53-acre site was operated as a railroad tie creosoting facility for four decades until it ceased operation in 1957. In the mid-1960s, a 137-property residential community was built on top of a portion of the site; several of the residences were located above waste lagoons full of concentrated creosote sludge. The problem did not emerge for 30 years. In 1996, a resident notified the New Jersey Department of Environmental Protection of a discharge of an unknown liquid from a sump. A year later, further evidence of contamination surfaced. A black tar-like substance was found in the soil during excavation of a sewer pipe in the area. Soil sampling by EPA revealed significant levels of polycyclic aromatic hydrocarbons at a number of residential properties. A record of decision, signed in the fall of 1999, called for the buyout and relocation of some 16 property owners. Expected costs of the Fund-lead remedial action range from $50 million to $100 million.[26]

The barrel-scraping effect may be more illusory than real if one considers two other channels by which expensive sites may end up on the NPL. First, assessment activities in other federal or state regulatory programs may in the future lead to the identification of new and potentially expensive NPL sites. In several instances, EPA regions and the states are undertaking a more targeted approach to site discovery to determine and assess possible contamination risks from former coal gasification plants and phosphate manufacturers. In a number of regions, EPA and state regulators are coordinating activities to identify potential sites as sources of releases that are contaminating municipal water supplies.[27] Public water systems in violation of drinking water standards under the Safe Drinking Water Act are the initial targets.[28] In several regions, EPA staff work closely with their state counterparts to devise a "watch list" of operating RCRA facilities that for various reasons could become NPL sites. Moreover, there are looming questions about how regulators will deal with contaminated sites as part of the total maximum daily load (TMDL) program under the Clean Water Act; such sites could be very expensive to address.*

Second, recent trends suggest that EPA may now be listing more mega sites than in the past. In FY 2000 and in the first quarter of FY 2001, EPA added five mega sites to the NPL: three mining sites (Leviathan Mine in California, Midnite Mine in Washington, and Gilt Edge Mine in South Dakota), one contaminated sediment site (Portland Harbor in Oregon), and one oil refining site (Indian Refinery–Texaco Lawrenceville in Illinois). Each of these sites is expected to cost more than $50 million to clean up. In addition, two mega sites were proposed to the NPL in the first quarter of FY 2001: Lower Duwamish Waterway, a contaminated sediment site in Washington, and the Del Amo site, a former synthetic rubber manufacturing facility in California. Adding more mining or contaminated sediment sites to the NPL would drive future Superfund costs up, but for a number of reasons—political, regulatory, and economic—it is highly uncertain how many of these sites, many of which would almost certainly be mega sites, will end up on the NPL and, if listed, whether they will be Fund-lead or PRP-lead.

Discussions with senior EPA managers made clear that with current funding levels, they cannot list many Fund-lead contaminated sediment or mining mega sites without "breaking the bank." Thus, the listing of mega sites is to some extent a result of how much money is available. A key question Congress and senior EPA management should address is whether these sites should be cleaned up as NPL sites. Currently, sites that need cleanup are not being addressed because of funding concerns. To illustrate this point and to describe more thoroughly why there is such uncertainty about mega sites in general, we discuss the particular cases of contaminated sediment and mining sites.

---

*Section 303(d) of the Clean Water Act requires states to identify waters that are not in compliance with water quality standards, establish priorities, and implement improvements. A TMDL is essentially a pollution budget. It specifies the amount of a particular pollutant that may be present in a water body, allocates allowable pollutant loads among sources, and provides the basis for attaining or maintaining water quality standards. On July 11, 2000, the final TMDL rule was signed and will become effective October 1, 2001.

Contaminated sediment sites

In the last decade contaminated sediments have become recognized as a major source of persistent and toxic pollutants in rivers, estuaries, and lakes.[29] Although there has not been a national assessment of the extent and severity of contaminated sediments, a recent screening assessment by EPA concluded that approximately 10% of the sediment underlying U.S. surface water is contaminated with pollutants, such as heavy metals, pesticides, polychlorinated biphenyls (PCBs), and dioxins, that pose potential risks to fish and wildlife and human health.[30] In addition, the Agency has identified 96 watersheds of "probable concern" for sediment contamination, including coastal estuaries, inland waterways, harbors, and urban rivers.[31]

Predicting how many contaminated sediment sites will be added to the NPL in the future is a complicated matter. First, the remediation of contaminated sediments poses difficult regulatory and management challenges. In the underwater environment, a cleanup strategy for contaminants in sediment layers may require treating large volumes of low contaminant concentrations, which if improperly handled can remobilize the contaminants and disperse them widely. Dredging and treatment of contaminated sediments may therefore pose unacceptable short-term risks and disturb sensitive natural environments, while less expensive remedies, such as *in situ* technologies or natural recovery, may under certain conditions not adequately address long-term risks to ecosystems and human health. Second, in many cases sediments have been contaminated by pollutants, such as DDT, PCBs, or agricultural runoff, that were discharged decades ago and have accumulated in the silt and mud of river and lake bottoms. In such instances, it may be difficult for EPA or a state program to identify a responsible party to pay for cleanup or to assign liability to parties without getting involved in a protracted legal case. Figuring out who is responsible is difficult in situations where both point and non-point sources of pollution upstream may contribute to the contamination of sediments downstream, say in a port community. As one EPA regional official told us, "…citizens want action, they want us to do a site assessment, but what is the size of the site and what is the background—two miles or two hundred miles—and how do you stop PRPs from suing everyone up and down the river?"[32] Furthermore, if the sources of contamination are to be addressed at the watershed level, how should costs for dredging, transporting, and treating contaminated sediments be allocated across jurisdictional boundaries?

Mining sites

There are an estimated 200,000 inactive or abandoned hardrock mines* in the United States, of which several thousand may pose significant threats to human health and the environment.[33] The environmental impacts of hardrock mining operations are extensive and well-documented: mine tailings, tailing ponds, and waste rock piles can release arsenic, selenium, cyanide, acid drainage, and other chemicals used in mining processes. Contaminants can be chemically leached or physically mobilized from these waste sources and transported to groundwater and distant surface water bodies. According to a recent National Research Council study, "Sediments and down-gradient soils may become contaminated and act as a secondary source of contamination to surface water and groundwater. Unless the waste materials are removed or effectively isolated from the environment by capping and/or lining, releases of contaminants, especially metals, can continue for long periods of time."[34]

From the inception of the Superfund program through FY 1999, 42 mining sites had been listed on the NPL, of which 17 were mega sites. Given the large number of abandoned and inactive mines that pose a serious threat to public health and the environment, why are more mining sites not on the NPL? There are a number of reasons. First, for abandoned mines, ownership can go back more than 100 years and include dozens of private and public entities; the inherent complexity of mineral and surface titles means that identifying PRPs is difficult and expensive. Second, even if identified through a laborious title search, a PRP may not have the financial resources to reclaim the mine, in which case federal and state agencies may have to pay for the cleanup. Indeed, the average cost to clean up a nonmega mining site on the NPL is about $22 million, more than double the average for most other nonmega site types, such as chemical manufacturing and recycling sites,† and the cleanup costs of mega mining sites can be exorbitant. At the Bunker Hill Mining and Metallurgical site in Idaho (made final on the NPL in September 1983), for example, EPA had spent approximately $212 million for various cleanup and management support activities through FY 2000.[35] Third, mining site remediation may require costly, long-term treatment of site dis-

---

*"Hardrock" mines include mining for gold, silver, copper, molybdenum, and other metals, but not coal mining.

†Estimates for nonmega sites include only extramural costs for RI/FSs, RDs, and RAs, and assume 1.6 operable units per site.

charges, and not unexpectedly, states may resist listing Fund-lead mining sites if they are required to pay both their 10% share of remedial action costs and 100% of what could be perpetual operation and maintenance costs.[36] More broadly, mining interests traditionally have wielded strong political influence in state capitals, and this pressure has tended to discourage the listing of mining sites.[37]

State regulators require financial assurance—typically a reclamation bond—from mining companies to assure that mines are returned to their original condition even if the operator goes bankrupt or is unable to finance post-closure remediation. Indeed, reclamation bonds have been used to help states pay for their contribution to cleanup costs for mining sites on the NPL. According to a recent comprehensive study of state reclamation bonding practices in the western United States, however, there is considerable variation among the states in closure cost estimates, ranging from $1,000 to $20,000 per acre.[38] Perhaps more importantly, at least for our purpose of estimating the future cost to the federal government, the study found that the financial assurances from mining companies were not sufficient to restore land and water resources damaged by mining operations, and that western states could face unfunded reclamation bonding liability exceeding $1 billion.[39]

If reclamation bonds do not provide sufficient financing for post-closure cleanup, federal and state regulatory agencies could face staggering costs to clean up mines that are abandoned or have no financially viable PRPs. Given federal and state superfund budgets, how likely is it that mines with no viable PRPs will be listed? Under what conditions would EPA—with state concurrence—list expensive, Fund-lead sites? One condition, perhaps, is when acid drainage or other contaminants, such as heavy metals, from mining sites contaminate already scarce drinking water supplies. With increasing numbers of people moving to the West and with more second-home construction in rural areas, this could become a more salient issue. A second condition relates to TMDLs: in some cases it will be difficult if not impossible to achieve water quality standards under section 303 of the Clean Water Act because of acid drainage and the leaching of mine wastes. The processes of acid generation and acid drainage are not well enough understood, and acid drainage has sometimes occurred at mines even when it was not predicted.[40]

The number of mega sites listed on the NPL in the next decade, especially those that are Fund-lead, is likely to be a key driver of long-term Superfund costs. And yet, as the examples of contaminated sediment and mining sites illustrate, decisions to list mega sites on the NPL will be contested in ways difficult to predict. The number of mega sites listed in the future is likely to be a function of funding availability, market conditions, political leverage, technological advances, and demographic trends.

Variation in State Capabilities

Since CERCLA was amended in 1986, states have come to play an increasingly important role in site cleanups. Most states have enacted their own site cleanup laws, known as state superfunds, which give state programs statutory authority to undertake their own cleanups and emergency response actions—paid for with state rather than federal funds—as well as the authority to compel responsible parties to conduct or pay for site remediation. More recently, the majority of states have established voluntary cleanup programs as an additional regulatory tool to encourage private parties to conduct cleanups of contaminated properties in the absence of state enforcement authority. States have completed cleanup of thousands of contaminated sites in both their enforcement-lead and voluntary cleanup programs.[41]

As noted earlier, an ASTSWMO survey of state hazardous waste officials reported that from 1993 through 1997, state programs addressed some 27,000 contaminated sites. A closer look at the survey data suggests that the majority of the sites addressed by state programs were less contaminated and less costly than a typical NPL site. For example, the survey calculated the duration of response actions for sites where removal or remedial actions were completed between January 1993 and October 1997. It found that removal or remedial actions at 76% of the 1,698 sites for which data were available were completed in less than 12 months, and at only 4% of sites did the duration exceed four years. Of the 1,191 sites for which cleanup cost data are available, the average cost per site was about $600,000, much less than the cost of an average NPL cleanup.[42]

The ASTSWMO survey results are not, unfortunately, disaggregated by site type or cost and cannot help us determine how many states programs are addressing sites that could otherwise be included on the NPL. Studies by GAO, however, suggest that a number of states have developed both the technical capabilities and the enforcement capacity to clean up complex, NPL-caliber sites,[43] a finding supported by our interviews with nine state officials. Most state officials told us that their programs had the capacity and the enforcement authority to pursue recalcitrant PRPs as well as the ability to reach settlements with large numbers of PRPs. In addition, they believed their programs had the technical expertise to clean up sites with severe contamination problems and could take on expensive remedies if the PRP funded the cleanup. However, state officials indicated that sites requiring expensive remedies and lacking financially viable PRPs to pay for cleanup were typically referred to EPA.

The future size of the NPL, then, will be influenced by the resources states have to finance their own fund-lead cleanups—that is, the money state agencies have to pay for cleanup at sites without financially viable and cooperative PRPs. According to a study by the Environmental Law Institute, expenditures from state hazardous waste cleanup funds in 1997 ranged from nearly $160 million in New York to less than $1 million in each of 16 states.* Twenty-seven states spent less than $5 million, and the median expenditure from state cleanup funds of the 44 states participating in the survey amounted to $2.27 million, considerably less than the average cleanup cost of a nonmega NPL site.[44]

A survey of 44 state hazardous waste officials by GAO supports the point that state capacity to take the lead on cleanups when there is no viable PRP varies widely. As Table 5-2 shows, roughly one quarter of state respondents rated their states' financial capability to clean up NPL-eligible sites excellent or good. These states had 15% of the 1,789 sites awaiting an NPL decision in CERCLIS in 1998. By contrast, more than half said their states' financial capacity was fair to very poor; this group of states accounted for 77% of NPL-eligible sites.

A state's capacity to clean up contaminated sites depends on many factors, including staffing levels, technical expertise, liability provisions in the state's statutes, and the adequacy of enforcement tools. Perhaps most importantly, it reflects the commitment each state makes to fund cleanup programs. Currently, state programs rely on a mix of funding sources, including appropriations from legislatures, bond issues, and taxes on hazardous substances.† Each of these funding mechanisms, however, has potential drawbacks and makes predicting a state's capacity to deal with orphan sites in the decade ahead more uncertain. For example, the level of state appropriations for a cleanup program could depend on the amount of state revenues available and be subject to political maneuvering; a bond issue typically requires a ballot proposal

---

*States were asked to report only money spent during the year, not money obligated.

†EPA provides states with Superfund core program grants to support program infrastructure, capacity building, and voluntary cleanup program development. In the past few years, these grants have averaged $16 million to $18 million a year. These funds are in addition to funds EPA gives to states for their activities at NPL sites.

Table 5-2. State Financial Capability and NPL Eligible Sites

State's financial capability to clean up potential eligible sites	Number of states	NPL eligible sites in states	
		Number	Percentage
Excellent	6	117	7%
Good	5	142	8%
Fair	7	666	38%
Poor	12	521	30%
Very poor	11	173	10%
No response	9	141	8%
Total	**50**	**1,760**	**100%**

Note: Does not include 29 eligible sites in the District of Columbia, Puerto Rico, and the Northern Mariana Islands or under the jurisdiction of the Navajo Nation, for which state financial capability data were not available. Totals may not add exactly because of rounding.

Source: Based on GAO interviews with state hazardous waste officials, reported in GAO, *Hazardous Waste: Unaddressed Risks at Many Potential Superfund Sites*, November 1998, p. 68–75.

and voter approval, which may be politically contentious; and although taxes on hazardous substances may provide a state program with adequate financial resources, as we know from the federal Superfund program, taxing authorities can expire and a new round of taxes can be difficult to impose.

One of the most significant findings from our interviews with EPA regional and state superfund managers is that state programs do not have adequate financial resources to undertake many Fund-lead cleanups of the magnitude typical at NPL sites. We asked each state official the extent to which the state superfund program had the funds to clean up orphan sites and sites with recalcitrant or nonviable PRPs. Only the more well-funded states in our sample claimed the ability to undertake such cleanups.

As is evident from Table 5-3, the amount of funding available to state programs to pay for state-lead sites varies considerably. The superfund program in New York State, for example, expended approximately $193 million on state-lead site remediation in its FY 2000–2001. This level of funding, however, is unlikely to continue.[45] Since 1986, the New York superfund program has been financed through a $1.1 billion bond act, the Environmental Quality Bond Act, but the funds were fully allocated at the end of March 2001. Proposed legislation to finance the state superfund program is based not on a long-term debt program, such as the 1986 bond act, but rather on a pay-as-

Table 5-3. Available Funding for State-Lead Cleanup

State	FY 2000 resources
California	$8 million
Florida	Subject to annual appropriations
Massachusetts	$3–5 million
Missouri	Capable of funding only small emergency response cleanups
New Jersey	$20–30 million (should stay in this range for next six years)
New York	$193 million
Ohio	Capable of funding only small emergency response cleanups
Pennsylvania	$45 million
Texas	$24 million

Source: RFF state interviews, July 2000.

you-go approach that requires the state legislature to allocate annual revenues to cover the annual cost of the program. According to officials in the New York Department of Environmental Conservation, the state's superfund program would receive approximately $90 million a year.[46] Apart from New York's program, even mature state superfund programs, such as those in California, Massachusetts, New Jersey, Pennsylvania, and Texas, have limited funds to clean up orphan sites, ranging from $3 million to $45 million on an annual basis for state-funded cleanups. Ohio and smaller state programs, such as Missouri's, have insufficient resources to clean up orphan sites and can fund only small emergency cleanups.

With limited resources for addressing orphan sites, state-funded cleanups rarely exceed $5 million for a single site, according to the officials we interviewed. For example, since 1997 the superfund program within the Pennsylvania Department of Environmental Protection, perhaps one of the strongest programs in the nation, has financed six cleanups whose cost exceeded $2 million each. The most expensive site Pennsylvania addressed cost $11 million, slightly less than the average cost of a nonmega NPL cleanup.[47] For the smaller states, remedial costs of $1 million typically exceed the programs' financial resources. As one state manager told us, "The state hazardous remedial fund is only capable of funding small emergency response cleanups, the state's 10% share of remedial actions, and 100% share of operations and maintenance at Fund-lead NPL sites."[48]

Limited funding will restrict the number of orphan sites and sites without a cooperating PRP a state program can take on. And without adequate and assured funding, sufficient staff, or the financial resources to address complex sites and recover costs from recalcitrant PRPs, it is unlikely that a state enforcement program will be seen by responsible parties as a credible threat. A state may use the threat of referral to EPA to encourage private parties to negotiate enforceable agreements or voluntary agreements with the state. The difficult question—for estimating the future size of the NPL—is how to quantify the effect that a viable threat of NPL listing has on PRP behavior and the ability of states to negotiate cleanups with PRPs in voluntary cleanup programs or state enforcement programs. Arguably, a credible Superfund program will encourage more PRPs to negotiate with states and to opt for cleanup in state enforcement or voluntary cleanup programs or to sign a consent decree with EPA and conduct the cleanup as an NPL equivalent site. Ironically, a strong federal program, one with adequate resources and enforcement capacity, is likely to reduce the number of sites on the NPL in the future.

Developing Estimates of Future Listings

Our estimates of the future size and cost of the NPL necessarily reflect a mix of quantitative analysis as well as expert opinion and professional judgement. In developing these estimates we attempt to balance recent programmatic trends, such as listing rates to the NPL and changes in the proportion of Fund-lead sites, with the views of EPA and state superfund managers. In this section, we use evidence from past trends and the result of interviews with senior EPA and state superfund managers to develop our estimates of the number and composition of sites added to the NPL from FY 2000 through FY 2009.

RFF interviewed all 10 of EPA's regional Superfund division directors as well as 9 state hazardous waste officials to learn more about the factors that influence site listing on the NPL; the kinds of sites that are likely to come onto the NPL in the future; how many sites EPA and the states believe will be listed in the next decade, focusing primarily on the next two to five years; and the likely cost of remedial action at these sites.* An important caution to note

---

*A copy of the interview questions and the criteria for selecting states to participate in the survey can be found in Appendix E; the states that participated in the interviews were California, Florida, Massachusetts, Missouri, New Jersey, New York, Ohio, Pennsylvania, and Texas.

before we discuss the results of our interviews is that most regions and states could identify sites likely to be listed on the NPL only one or two years into the future and were either unable or unwilling to speculate on the size of the NPL beyond five years. In some instances, EPA regional staff and state program managers did not want to identify potential sites where negotiations with responsible parties were in progress. Others preferred not to disclose site-specific information for political or administrative reasons. Superfund program mangers were also hesitant to estimate the number of potential NPL sites in their region or state: The lack of proactive site discovery programs meant it was not possible to estimate the universe of contaminated sites in their regions. And, of the sites that have already been discovered and are already in CERCLIS or state inventories, many were in the early stages of investigation or had not been assessed, making it difficult to draw conclusions about their future disposition. According to most regional managers, there are also likely to be "pop-up" sites that are impossible to predict. Many of the program staff in the regional offices and states believe that although it is important to ask questions about the number and character of future NPL sites, it is perhaps futile to try to estimate their number beyond one or two years in view of shifting political currents, funding availability, and economic changes. Despite these caveats, we were able to identify some common themes and trends from our interviews.

In the next two to five years the number of sites added to the NPL is likely to increase.

In each interview with EPA regional managers, we asked whether they expected the rate of NPL listings from their regions to remain fairly constant, to increase, or to decrease in the next two to five years. Nine of the 10 EPA regions anticipated that listing rates would increase. One region expected to continue listing 7 to 8 sites per year. When we aggregated their responses, the regions predicted an increase in the number of NPL sites ranging from 49 to 63 sites a year. Explanations for this increase fell into three categories. First, according to some of our respondents, this increase reflects pent-up demand in the system. Now that 57% of sites listed on the NPL have reached construction completion,* program managers can address sites at the head of the

---

*Based on the number of construction-complete sites as of the end of FY 2000.

queue. Second, some state and EPA regional staff see this increase as part of a thaw in EPA–state relations, with states having more confidence in EPA because of the Agency's more streamlined listing process and other reforms. States are thus more willing to list sites on the NPL. Some state respondents, while not disputing the increase in future listings, suggest that EPA is now putting more pressure on them to add sites to the NPL, and it is this pressure rather than EPA's reforms that will lead to increased listings. Lastly, in FY 1992, EPA regions implemented the Superfund Accelerated Cleanup Model. Under this policy, removal actions in some regions were sometimes used to remediate sites, and as a result, the sites did not need to be put on the NPL. Recently, this policy has changed, and non-time-critical removal actions are being discouraged as the sole basis for remediating sites not already on the NPL. As a result, some state and regional managers expect NPL listings to increase in the future.

Although EPA regional staff and state superfund managers agreed that more sites would likely be listed on the NPL in the future, the estimate of 49 to 63 new sites each year may be too high. Superfund managers focused on the supply of potential NPL sites, but their estimates, in our view, did not give adequate weight to the political pressures that are likely to limit the ability and willingness of EPA to add sites to the NPL. For example, as noted earlier, EPA and the states have deferred final decision on some 2,500 sites in CERCLIS classified as NPL-eligible, that is, sites that scored 28.5 or above on the HRS. Indeed, nearly 75% of these sites were placed in CERCLIS before 1990. As one commentator put it, "There appears to be no relationship between how long a site has been awaiting an NPL decision and the likelihood that some cleanup has occurred during that time. It is uncertain when and how most of these sites will, ultimately, be cleaned up...."[49] Since the GAO report was issued in November 1998, EPA regions and the states have been discussing work-share arrangements for the NPL-eligible sites in CERCLIS to determine whether the federal or state programs will take the lead at these sites. This is a welcome initiative, but there remain nearly 1,100 sites for which the lead has yet to be determined. If EPA and the states are not able or willing to divide responsibility for cleaning up these sites, arguably the large influx of sites onto the NPL anticipated by regional EPA managers may be overstated.

Another complicating factor in establishing a listing rate is the governor's concurrence policy, noted earlier. This requirement, which was included in a 1996 appropriations bill, prohibited EPA from listing or proposing a site for listing unless the governor provided a written request ("concurrence") that the site be added to the NPL. Although appropriations bills in subsequent years have not contained such a provision, EPA established a policy of coordinating with the states in NPL listing decisions. Typically, the EPA regional administrator asks the governor to explain the position of the state on sites the region is considering for NPL listing.[50] Although EPA can now propose sites to the NPL without the governor's concurrence, since 1996 the Agency has done so only once, for the Fox River NRDA/PCB Releases site in Wisconsin.* No sites have been made final on the NPL over a governor's objection. In addition, the governor's concurrence policy may have a more subtle effect. For example, to what extent will EPA develop an HRS package at a site for which the Agency is unlikely to get the governor's concurrence? Or if an HRS package is in hand, what effect will a state's position against listing have on EPA's decision?

Finally, the steep increase in listing rates predicted by EPA regional managers does not accord with recent trends in the program. Figure 5-2 shows the number of final sites listed on the NPL from FY 1983 through FY 2000. The early years of the program were marked by great fluctuations, with hundreds of sites listed in one year and few if any sites listed the following year. Since FY 1993, however, the early exuberance and unpredictability of the program's listing rates have not continued; as the program has matured and as more administrative reforms have been implemented, the number of sites added to the NPL has become more consistent. Since 1996, even with the governor's concurrence policy in effect, the number of nonfederal sites added to the NPL has averaged 23 final listings per year. Recently, there has been a marked increase, with more sites listed in each of the past two years (FY 1999 and FY 2000) than in any year since 1990.

In our model of future costs, we assume that the average number of sites (23) added annually to the NPL since FY 1996 can serve as a baseline for EPA's willingness and ability to list sites in the coming years. Given the degree of uncertainty in any forecast of the future size of the NPL, we posit a *range* of likely future NPL additions rather than a single estimate. Our interviews strongly suggest an increase in listings; based on this we assume 49 sites per year is the likely high end of the range. Our base case, which we believe to be the most likely, assumes 35 new sites per year.

---

* The Fox River site was proposed to the NPL on July 28, 1998, and has not been made final.

Figure 5-2. Sites Listed as Final on NPL, FY 1983–FY 2000

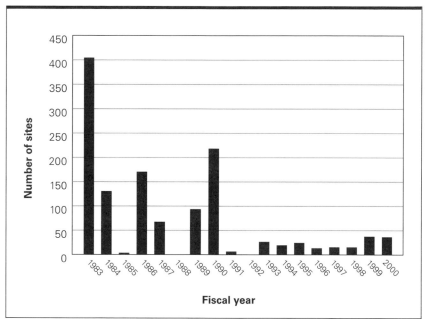

Note: Based on nonfederal sites.
Source: RFF site-level dataset. FY 2000 data provided to RFF by EPA, October 2000.

The sites added to the NPL in the future are likely to be more costly and complex.

The majority of EPA regional and state managers anticipate listing a higher proportion of groundwater contamination sites, contaminated sediment sites, mining sites, and smelter sites on the NPL than in recent years. Although in theory many of these sites can be handled under state cleanup programs, such sites commonly have multiple sources of contamination, enforcement problems related to recalcitrant or insolvent PRPs, and complicated liability issues. According to our interviews, most state managers are confident their programs can handle technically complex sites but prefer to list sites on the NPL when liability is beyond the reach of the state, thus requiring EPA's enforcement authority, and when they do not have adequate funds to deal with recalcitrant parties or orphan sites.

Our long-range estimates of Superfund program costs are sensitive not only to the number of sites we assume will be placed on the NPL but also to the number of mega sites. The 114 mega sites on the NPL at the end of FY 2000 accounted for 9% of all Superfund sites. As Figure 5-3 shows, the rate at which mega sites have been added to the NPL has varied considerably over the course of the Superfund program. Nearly 75% of all mega sites were listed in the early years of the program. Since FY 1987, on average, 2.2 mega sites per year have gone final on the NPL. In the past five years, from FY 1996 through FY 2000, the listing rate for mega sites has dropped to an average of 1 site per year. This would suggest an overall downward trend in mega site listings, particularly when the additions to the NPL from FY 1983 through FY 1986 are compared with those from FY 1996 through FY 2000. In FY 1999 through the first quarter of FY 2001, however, 8 mega sites were listed as final on the NPL.

In our interviews with EPA regional staff, we asked whether the region had a list of specific sites it would like to add to the NPL and how many of these sites

Figure 5-3. Mega Sites Listed as Final on NPL, FY 1983–FY 2000

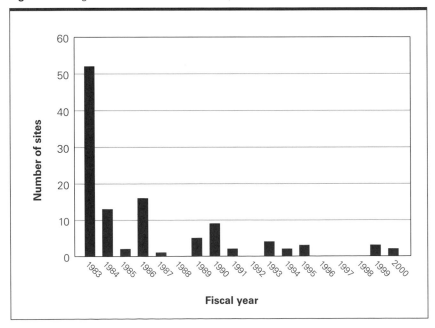

Note: Based on nonfederal sites.
Source: RFF site-level dataset. FY 2000 data provided to RFF by EPA, October 2000.

were likely to be mega sites. For enforcement reasons, most EPA regions did not provide site-specific details of potential mega sites but did indicate that they were aware of a total of 10 to 15 potential mega sites that could be listed in the next two to five years, including contaminated sediment sites, mining sites, former transportation facilities, landfills, and manufacturing sites. EPA regional managers indicated that they believed responsible parties were likely to pay for remedial actions at roughly 80% of these sites. There was considerable uncertainty, however, about the number of mega sites that would ultimately be placed on the NPL. Political resistance to listing by state governments, the controversies surrounding dredging of contaminated sediments, the exorbitant cost of cleaning up these sites, and the other options for addressing them—such as addressing the site as an NPL equivalent—make estimating the number of future mega sites to be added to the NPL highly uncertain.

From our interviews with EPA regional Superfund managers, it appears that the barrel-scraping effect we noted earlier can be discounted. The Agency is already aware of 10 to 15 potential mega sites, and both EPA regional staff and state superfund officials believe that in the next 5 to 10 years additional mega sites will, as one official put it, "come out of nowhere." Based on our interviews and recent trends, we assume that on average one to three mega sites will be added to the NPL each year from FY 2000 through FY 2009.

EPA regions anticipate an increase in the proportion of Fund-lead actions at NPL sites through at least FY 2005.

From FY 1996 through FY 2000, responsible parties paid the cost of 72% of remedial actions at NPL sites, with the Fund accounting for the remaining 28% of remedial action starts. In our interviews we asked EPA regional managers whether they expected the Fund-lead share of remedial actions to increase or decrease in the coming years. Based on planning data for remedial action starts as well as projections through FY 2005, nine regions responded that more cleanups at the remedial action stage would be Fund-lead in the future; in one region, the split between Fund-lead and PRP-lead cleanups was predicted to remain the same. In total, EPA regional managers projected that responsible parties would pay the cost of cleanup at 57% of the remedial action starts during the next five years, with Fund-lead remedial actions increasing from 28% to 43%.[51] The primary reason given for this increase in Fund-lead cleanups, by both EPA regional staff and state superfund directors, is that states are now addressing the majority of single-party sites and sites with cooperative responsible parties. These sites are being cleaned up in state voluntary programs or enforcement programs, leaving EPA the orphan sites and sites with recalcitrant PRPs—that is, the sites more likely to have Fund-lead actions.

The NPL from FY 2000 to FY 2009: Three Scenarios

To develop estimates of the future cost of sites added to the NPL from FY 2000 through FY 2009, we had to make assumptions about each of the three cost drivers: the number of nonfederal sites added to the NPL each year, the number of mega sites added each year, and the proportion of remedial pipeline actions that will be Fund-lead. Based on our research, we developed three scenarios regarding the number and type of sites that will become final on the NPL from FY 2000 through FY 2009. In all three scenarios we use actual data for NPL final listings in FY 2000: 36 new NPL sites, including 2 mega sites.* The three scenarios described below, and summarized in Table 5-4, are based on variations in each of the three factors.

In the low case, we assume that future listing rates and the incidence of mega sites will look very much as they have during the program's past five years. We assume that 23 sites, 1 of which is a mega site, will be added to the NPL each year from FY 2001 through FY 2009. In this scenario, we project that the percentage of Fund-lead actions will be consistent with recent history (FY 1996 through FY 1999) at both mega and nonmega sites.

In the base-case scenario—which we believe is our best estimate of the number and composition of sites likely to be listed on the NPL in the future—we assume that 35 sites will be added to the NPL each year from FY 2001 through FY 2009. This number is 50% higher than the average annual NPL listing rate from FY 1996 through FY 2000. However, EPA and state superfund managers reported that since the mid-1990s, site listings have been "artificially constrained" as federal and state resources were allocated to the back end of the CERCLIS pipeline and to achieving construction-complete status at current NPL sites. In the view of many state and regional officials, the pendulum is swinging back, and they expect to list more sites in the next two to five years—a trend already evident in FY 1999 and FY 2000. Of these annual additions, we assume that two sites will be mega sites, based in large part on our interviews with regional EPA staff and the recent upswing in the

---

*There were 37 sites listed on the NPL in FY 2000, but one of these sites, Georgia-Pacific Corp. in North Carolina, was withdrawn from the NPL on September 28, 2000.

Table 5-4. RFF Future NPL Scenarios, FY 2001–FY 2009

	Sites listed each year		Ratio of Fund-lead to PRP-lead actions	
	Total	**Mega**	**At nonmega sites**	**At mega sites**
Low case	23	1	Based on FY 1996–FY 1999	Based on FY 1996–FY 1999
Base case	35	2	Increase in Fund-lead share	Based on FY 1996–FY 1999
High case	49	3	Increase in Fund-lead share	Increase in Fund-lead share

listing of mega sites noted earlier. We also assume that the proportion of Fund-lead actions at mega sites will be consistent with recent history (FY 1996 through FY 1999) as well as with the information we collected in our interviews with EPA regional division directors. For nonmega sites, we assume that the percentage of Fund-lead actions will be higher than in recent history as states address more sites with viable PRPs in their enforcement or voluntary cleanup programs and increasingly refer sites to the NPL when there are recalcitrant or insolvent PRPs and Trust Fund dollars are needed.

Lastly, in the high case, we posit a sharp increase in the number of sites added to the NPL, slightly more than double the average of the last five years. This estimate reflects what many regional and state superfund managers consider the need for cleanup. In the words of one EPA regional official, this is the number EPA and the states could list if they had "gotten up the gumption." Here we assume that 49 sites will be added to the NPL each year from FY 2001 through FY 2009, 3 of which will be mega sites. We also assume that the proportion of Fund-lead actions will increase at both mega sites and nonmega sites. It is worth noting that this number, although higher than in recent years, is still lower than the high estimate provided by EPA regional managers in our interviews.*

---

* See Chapter 7 and Appendix G for a more detailed description of our approach to estimating the cost of future NPL sites under three scenarios.

Chapter Six

Superfund Support Activities and Programs

As discussed in Chapter 1, we divided the Superfund program into six elements in order to estimate its future cost. In the preceding chapters, we focused on the two core elements of the Superfund program that account for the direct costs of cleaning up sites—the removal program and the remedial program—while also addressing the cost of U.S. Environmental Protection Agency (EPA) site assessment efforts as part of the discussion of how sites are listed on the National Priorities List (NPL). In this chapter, we turn our attention to the other three major elements of the Superfund program: (1) program staff, management, and support, (2) program administration, and (3) other programs and agencies, which we loosely group together under the umbrella of Superfund support activities and programs. In FY 1999 these activities accounted for just over a third of the total dollars ($540 million of $1.54 billion) spent by EPA to implement the Superfund program.

The cost of many Superfund support activities and programs are not charged directly to sites. For this reason, these activities are often termed nonsite activities, and their associated costs are frequently referred to as nonsite costs. However, many of these activities are, in fact, necessary components of site cleanup efforts. For example, included in this category are rent, staff payroll and benefits, and policy development. Although it is certainly reasonable to question whether EPA is spending the appropriate amount on these activities, it is simply not possible to have a national Superfund program without them.

One reason we discuss these three program elements together is that we were not able to confidently assign the costs of such activities to the removal and remedial programs. Additionally, we were not able to determine what proportion of these costs is truly "required" for a successful site cleanup program. Our inability to do so stems in large part from the manner in which these costs are coded and categorized in the Agency's internal financial system, the Integrated Financial Management System (IFMS), which is our source of expenditure data.*

We begin this chapter with a brief overview of how we assigned costs to the three program elements. We then describe each element, the activities it encompasses, and what it cost EPA to carry out these activities in FY 1999. As part of this discussion we identify, where possible, the activities undertaken in direct support of the removal and remedial programs. We conclude this chapter with a discussion of cost elements that we believe deserve further analysis, such as the high percentage of regional staff time not billed to specific sites.

Assigning Costs to Superfund Support Activities and Programs

Ideally, one would be able to assign a specific proportion of the costs of staff, management, and support efforts and administrative functions to the removal and remedial programs. Unfortunately, the manner in which expenditures are coded in IFMS does not facilitate such an approach. In some cases similar costs are coded dissimilarly by different offices within the Agency; in other cases, it was extremely difficult to understand what kinds of activities were actually included in certain IFMS cost categories. Given these constraints, we decided instead to organize these activities into easily understandable categories and to identify uncertainties about these costs where appropriate.

The three program elements are as follows:

- *program staff, management, and support:* staff payroll and benefits for general management and support activities associated with the remedial and enforcement programs (such as the drafting of policy and guidance documents),† response action management (such as contract and information management), state programs and grants, laboratory analysis, and various EPA offices that provide policy, legal, and research support;
- *program administration:* rent, facilities operation and maintenance, equipment and supplies, activities carried out by the EPA Office of Administra-

---

*We did not evaluate EPA's indirect cost accounting methodology as part of our analysis of IFMS, and nothing in this report should be taken as either support or criticism of that methodology or any of its components.

†We were able to assign most of the costs of payroll and benefits to the removal program and to site assessment activities.

tion and Resources Management (OARM), regional administrative and resource management staff time, and activities performed within the EPA Office of the Chief Financial Officer (OCFO);* and

• *other programs and agencies:* Superfund-related work of offices within EPA, including the Technology Innovation Office (TIO), the Chemical Emergency Preparedness and Prevention Office (CEPPO), and the Office of the Inspector General (OIG), and other federal agencies, including the National Institute for Environmental Health Sciences (NIEHS) and the U.S. Coast Guard.

We use FY 1999 expenditure data to provide a baseline for how much money EPA spends on each activity. Although the total spent on each of these three elements is consistent with actual FY 1999 expenditures, in a number of cases we had to develop cost estimates for specific activities within each of these categories using EPA budget information (i.e., beginning-of-the-year budget allocation assumptions).† Thus, for example, IFMS expenditure data provided the total amount spent for regional payroll and miscellaneous administrative and contract expenses in the enforcement program, $79.1 million in FY 1999, but our allocation of this expenditure to two major types of activities—"maximizing PRP involvement" and cost recovery—was arrived at by applying the budget plan assumptions of a 75% to 25% ratio.

Program Staff, Management, and Support

The part of the Superfund program we categorize as program staff, management, and support represents a catchall category of activities implemented with Superfund monies. These activities include a wide range of efforts carried out in different EPA offices, regional staff time not charged specifically to site accounts (e.g., paid sick leave, federal holidays, vacation, staff meetings), the work of private contractors [e.g., to support Agency information management, such as the Comprehensive Environmental Response, Compensation,

---

*Some of the activities carried out by OCFO's Financial Services Division (and their associated regional counterparts) provide direct support to the cleanup programs, such as preparing Superfund cost recovery packages, attending site negotiations, providing depositions, and testifying at cost recovery trials.

†To develop these estimates, we worked with staff in EPA's Office of Emergency and Remedial Response, Office of Enforcement and Compliance Assurance, OARM, and OCFO.

and Liability Information System (CERCLIS) and regional laboratories], and a handful of other programs. In FY 1999, EPA spent $325.6 million, or 21% of its total Superfund expenditures, on these activities. Almost two-thirds of this amount ($210.1 million) was spent in the 10 EPA regional offices, with the balance ($115.5 million) spent at EPA headquarters. Table 6-1 shows the major components of these expenditures and the amounts spent in the 10 EPA regional offices and EPA headquarters.

Remedial Program Policy, Management, and Staff

The largest component of the costs in this category is remedial program policy, management, and staff, which in FY 1999 accounted for $113.9 million. More than a third of the total, or $42.5 million, was for costs associated with staff at EPA headquarters—primarily within the Office of Emergency and Remedial Response, which is responsible for developing national

Table 6-1. FY 1999 Expenditures for Program Staff, Management, and Support (Millions of Dollars)

	Regions	Headquarters	Total
Remedial program policy, management, and staff	71.4	42.5	**113.9**
Enforcement program policy, management, and staff	79.1	20.0	**99.1**
Response action management	37.5	—	**37.5**
State programs and grants	20.5	10.9	**31.4**
Laboratory support	—	16.6	**16.6**
Office of Solid Waste and Emergency Response front office	—	4.0	**4.0**
Office of Enforcement and Compliance Assurance front office and communications	—	2.3	**2.3**
National Enforcement Investigations Center	—	13.7	**13.7**
Office of Research and Development	—	4.0	**4.0**
Office of General Counsel	1.6	1.5	**3.1**
Total	**210.1**	**115.5**	**325.6**

Source: Based on data provided to RFF by EPA, August 2000.

Superfund policy. These general policy and management efforts include the following:

- developing technical policies and guidance for application in site-specific situations (e.g., presumptive remedies guidance);
- setting priorities for funding response actions by the National Prioritization Panel;
- providing headquarters policy support to regions, states, and tribes on site-specific technical issues;
- developing guidance documents for conducting investigations, risk assessments, evaluating and documenting remedy selections, and evaluating progress toward construction completion and post-construction completion activities;
- evaluating alternatives to NPL listings (e.g., NPL equivalent sites);
- developing policies and actions to enhance the capabilities of states and tribes to implement their superfund programs or to participate in the national program;
- providing assistance to improve community involvement at sites (e.g., technical assistance grant rules, environmental justice issues, and development of relocation guidance); and
- responding to congressional requests, preparing budgets, participating in technical and policy workgroups, and arranging national meetings.

The remaining $71.4 million associated with remedial program policy, management, and staff was spent in EPA's regional offices. The Agency spent $11.4 million for equipment, travel, general supplies and materials, training, and other miscellaneous goods and services. Roughly $60 million of these regional expenditures can be attributed directly to staff payroll and benefits. This is somewhat surprising, as one would assume that the majority of regional Superfund staff would, for the most part, be working directly on site-specific matters. In fact, of the total $208 million spent for payroll expenses from Superfund accounts in the 10 EPA regional offices in FY 1999, only 25% was charged to site-specific accounts.[1]

Certainly no one would expect that 100% of all regional EPA staff time be charged to site-related activities, as some proportion of these costs covers the time of senior management and administrative staff. Such costs relate not only to efforts to cleanup NPL sites, but also to the removal program and other regional initiatives. Additionally, staff-related costs for sick leave, paid vacation, and federal holidays are not charged to specific sites in IFMS. Although it is not clear what percentage of regional staff time *should* be charged to site-specific activities (most likely somewhere between 50% and 60%), it is unlikely that all $60 million in regional payroll and benefit expenditures not charged specifically to sites is for staff time related to management and other activities.

There are a number of possible explanations for why such a large proportion of regional staff time is not charged to specific Superfund site codes: Staff are not charging their time correctly (i.e., they are spending more time working on specific sites than is indicated by IFMS data), or there are more nonsite activities and responsibilities (for example, training, administrative responsibilities, participation in policy and guidance workshops) than one would expect. However, short of conducting extensive interviews or performing a "desk audit," we have no way to determine what percentage of the regional payroll and benefits expenditures is for program management costs, and what percentage is for the costs of other activities, or what these other activities include. More importantly, we cannot discern how much is necessary for site cleanup efforts under the remedial program. As a result, we include these costs in the program staff, management, and support program element.

Enforcement Program Policy, Management, and Staff

Also included in the program staff, management, and support category are EPA enforcement policy and management efforts. The picture here is similar to that of the remedial program discussed above. Of the $99.1 million spent on enforcement efforts not charged to specific sites as part of the removal or remedial (NPL and non-NPL) programs in FY 1999, $20 million (20%) was spent at EPA headquarters for a variety of policy and management activities. These activities include settlement policy development and implementation, accomplishment tracking and reporting, and information management. Enforcement staff in EPA headquarters are also involved in site- and case-specific assistance to EPA regional offices (another example of costs that could be charged directly to sites), enforcement case priority planning, and pre- and post-litigation support. The remaining $79.1 million spent in FY 1999 for enforcement efforts covered payroll and miscellaneous administrative and contract expenses in the 10 EPA regional offices not charged to specific sites. According to EPA, this $79.1 million was spent for activities that can be grouped into two categories: maximization of PRP involvement and cost recovery, for which EPA spent $59.3 million (75%) and $19.8 million (25%), respectively.[2]

Maximizing PRP involvement consists of a number of different activities to get parties responsible for contamination at sites to either fund or conduct cleanups, as well as implementing recent administrative reforms. Among these activities are identifying and negotiating with PRPs, issuing cleanup orders, determining allocations of orphan share offers, protecting *de minimis* and *de micromis* parties from third-party litigation,* and taking actions necessary to ensure that PRPs comply with the terms of consent decrees and administrative orders. EPA's cost recovery efforts include seeking reimbursement for the cost of site cleanup from responsible parties, as well as records management, cost documentation, alternative dispute resolution, litigation referrals, and case support.

As was the case for remedial program policy, management, and staff, a high percentage of payroll costs for enforcement-related activities in the regions is not charged to site accounts. Of the almost $70 million spent in FY 1999 in the 10 EPA regional offices for enforcement-related payroll, $53 million, or more than 75%, was not charged to specific sites. Again, some of these costs certainly are incurred for management and administrative activities and for staff benefits, but one would anticipate that most other enforcement efforts, whether for maximizing PRP involvement or for cost recovery, would be directly related (and charged to) to specific sites.

Response Action Management

A number of management and support activities constitute the category we call response action management. These activities include removal and remedial contracts that are not assigned to a specific site, support of contractor infrastructure, and contracts that support EPA information management, including those for regional records management centers and CERCLIS. In FY 1999, EPA's total expenditure for response action management was $37.5 million.

State Programs and Grants

Also included in program staff, management, and support is funding for state programs and grants, which collectively accounted for $31.4 million in FY 1999 expenditures. About two-thirds ($20.5 million) of the money spent on

---

De minimis parties are persons who, pursuant to section 122(g) of CERCLA, contributed hazardous substances to a facility, which are minimal in both volume and toxicity or other hazardous effects, relative to other hazardous substances at the site. *De micromis* parties are persons whose contribution is equal to or less than (1) 0.002% of the total volume of waste at the site or 110 gallons (such as two 55-gallon drums) or 200 pounds of materials containing hazardous substances, whichever is greater, or (2) 0.2% of total volume of waste at the site, where a contributor sent only municipal solid waste.

state programs and grants is for cooperative agreements between EPA regional offices and states for core program work that is not assigned to a specific site.* Core program grants enable the states to build and maintain capacity in their superfund programs as well as support the federal Superfund program. EPA headquarters allocated the balance of the money ($10.9 million) to states and tribes through cooperative agreements and grants for program management assistance and training, to state and tribal associations to conduct studies of the Superfund program, and to participate in federally funded state and tribal activities (such as conferences). Some funds also go to universities for technical outreach support to communities near contaminated sites.

Laboratory Support

EPA incurred $16.6 million in FY 1999 in support costs for laboratories that conduct analysis for specific sites and for contractors who provide performance evaluation samples, assess data, and schedule site work. Other support activities include development of lab sampling protocols and on-site laboratory audits for quality assurance.

Other Offices and Programs

Program staff, management, and support also encompasses several offices within EPA. These offices (and their associated Superfund costs in FY 1999 expenditures) include the following:

- Office of Solid Waste and Emergency Response (OSWER) front office ($4 million): management and coordination of Superfund planning, budgeting, analysis, and financial management services;
- Office of Enforcement and Compliance Assurance front office and external communications ($2.3 million): management and planning of Superfund-related enforcement activities, and public outreach and communication efforts;

*EPA also provides states with funds for site-specific activities at NPL sites, for example, to conduct remedial pipeline actions. For the purpose of our analyses, these funds are included as part of the costs of Fund-lead actions, since in terms of the overall mission of the Superfund program, they are indistinguishable from funds EPA provides to contractors or to the U.S. Army Corps of Engineers to implement Fund-lead cleanup activities.

- National Enforcement Investigations Center ($13.7 million): activities supporting CERCLA civil investigations, such as sampling, laboratory, and financial analysis;
- Office of Research and Development ($4 million): research activities related to the remediation of contaminated sites*; and
- Office of General Counsel ($3.1 million): provision of legal advice regarding Superfund to all parts of the Agency.

Program Administration

In addition to staff and contractor costs, the Superfund program, like any other governmental or nongovernmental program, incurs administrative costs, such as those for rent, maintenance of financial accounting systems, and contract management. In FY 1999 program administration costs incurred by EPA for the Superfund program totaled $114.1 million, of which about $44.5 million was spent in the EPA regional offices and about $69.6 million was spent in EPA headquarters.† None of these costs are billed site-specifically.

For simplicity, we have divided program administration costs into three categories:

- rent, facilities operation and maintenance, and equipment and supplies: rent for all EPA offices (headquarters and regional), utilities, security, property management, transportation services, and administrative equipment, supplies, and materials;
- activities carried out by OARM and regional administrative and resource management staff: contract and grant management, human resources management, information systems and services, and regional staff time and travel associated with administrative and resource management efforts; and
- activities performed within OCFO: management of EPA's financial systems (e.g., disbursement of payroll, processing of interagency agreements), coordination, oversight of the Superfund budget, and performance of annual

---

*This does not include research and development funded as part of the science and technology appropriations. For the past several years, Congress has designated more than $30 million of the annual Superfund appropriation for research and development; this money is transferred by law directly into the science and technology account and is not considered part of EPA's Superfund expenditures.

†All rent costs are charged to EPA headquarters accounts.

Table 6-2. FY 1999 Expenditures for Program Administration (Millions of Dollars)

	Regions	Headquarters	Total
Rent,[a] facilities operation and maintenance, and equipment and supplies	16.2	45.0	**61.2**
Office of Administration and Resources Management activities and regional administrative and resource management staff time	19.8	14.9	**34.7**
Office of the Chief Financial Officer activities	8.5	9.7	**18.2**
Total	**44.5**	**69.6**	**114.1**

[a]All the costs of rent (including those for EPA regional offices) are captured in the expenditures of EPA headquarters.

Source: Based on data provided to RFF by EPA, August 2000.

and strategic planning of the Superfund program in accordance with the Government Performance and Results Act.

Table 6-2 summarizes FY 1999 expenditures for the categories of program administration.

Although all the costs discussed in this chapter are based on IFMS expenditure data, EPA's derivation of program administration costs is slightly different. Many of the administrative costs are not specifically charged to the Superfund program as they are incurred; instead, costs are allocated to the program by formula. For example, EPA allocates a percentage of its rent costs to each program within the Agency based on the percentage of total work years—referred to as full-time equivalents (FTE)—in each program. Similarly, formulas are used to determine Superfund's share of financial management and human resources management costs. EPA uses a variety of approaches for determining what percentage of program administration costs should be paid for with Superfund monies. This method of determining the amount of administrative costs each program is responsible for is quite common for organizations (whether public, private, or nonprofit) that need to allocate such costs among several programs or offices.

Other Programs and Agencies

A number of EPA programs, both within and outside OSWER and in other federal agencies, are paid for with Superfund monies, but they either are not directly related to the removal and remedial cleanup programs or, if they are Superfund related, are "pass-through" funds—that is, their funding is determined directly by Congress. In FY 1999, $100.5 million, or almost 7% of EPA's total Superfund expenditures, was spent for these programs and agencies.* The programs and agencies and their associated costs are briefly described below.

TIO, in OSWER, accounted for about $8.1 million of EPA's Superfund costs in FY 1999. This office was created in 1990 to serve as an advocate for the use of new technologies—specifically, to increase the application of innovative treatment technologies and site characterization techniques of contaminated waste sites, soils, and groundwater.

CEPPO, also housed in OSWER, accounted for $7.6 million in Superfund costs in FY 1999. CEPPO uses Superfund resources to develop, implement, and coordinate efforts to prevent and prepare for chemical emergencies, respond to environmental crises, and inform the public about chemical hazards in the community. More recently, CEPPO has become involved in the federal government's counter-terrorism program.

Part of EPA's Office of the Inspector General (OIG) budget comes from Superfund resources as well. In FY 1999, $20 million from the Superfund program went to finance activities within this office. Among the Superfund-related duties of OIG is complying with legislative audit requirements, such as annual audits of the Trust Fund, response claims, and state cooperative agreements. Additionally, OIG performs mandatory financial audits of EPA's management of the Superfund program.

Two federal agencies receive resources from EPA as part of the Superfund program—NIEHS† and the Coast Guard. Together these agencies received $64.8 million in FY 1999: $60 million for NIEHS and $4.8 million for the Coast Guard, amounts specified by Congress in the appropriations bill that funded the Superfund program.

NIEHS administers two programs with its Superfund resources. The first is the Superfund basic research program, which was authorized by the 1986 Superfund Amendments and Reauthorization Act and is a university-based grant program that promotes the development of methods and technologies to detect hazardous substances in the environment and methods to assess the risks to human health presented by hazardous substances.[3] The second program is

---

*The dollars associated with these programs and agencies are based on the allocations included in EPA's FY 1999 operating plan, rather than on FY 1999 expenditures.

†Beginning in FY 2001, the Superfund-related work of NIEHS was funded through direct appropriation from Congress, rather than as part of EPA's Superfund appropriation.

the worker education and training program, which provides training for hazardous waste cleanup workers. The Coast Guard's Superfund-related work includes response to actual or potential releases of hazardous substances involving the coastal zone, Great Lakes, and designated inland river ports.[4]

The Department of Justice (DOJ) receives approximately $30 million each year in Superfund monies to cover the costs of enforcement activities. DOJ attorneys and support staff from the Environment and Natural Resources Division are responsible for conducting all civil judicial enforcement litigation under CERCLA, representing EPA and other federal agencies in Superfund-related bankruptcy proceedings, prosecuting criminal violations of the statute, conducting defensive and appellate litigation, negotiating settlements, resolving claims against small contributors to a site, and processing administrative settlements that require the Attorney General's concurrence.[5] All DOJ expenditures are ultimately charged to specific cases associated with specific sites, and thus we were able to link these expenditures directly to the removal and remedial programs.

In addition to NIEHS, the Coast Guard, and DOJ, a number of other federal agencies also conduct work that supports the Superfund program. According to EPA's FY 1999 operating plan, $76 million went to the Agency for Toxic Substances and Disease Registry, $2.45 million to the National Oceanic and Atmospheric Administration, $1.1 million to the Federal Emergency Management Agency, $1 million to the Department of the Interior, and $0.65 million to the Occupational Safety and Health Administration. Congress determines the level of funding for these agencies, and EPA simply transfers the congressionally mandated amount directly to the agencies at the beginning of the year. EPA does not track how these funds are used, and because the funding levels for these programs are set directly by Congress, we did not evaluate them in this study.

Conclusions

Like all organizations, public or private, the Superfund program incurs costs for a variety of mundane activities, such as rent, contract management, and accounting—all necessary components of its day-to-day operations. There are also costs for developing national policies and regulations, as well as for carrying out general managerial functions. It is easy to forget these program management and administration costs when considering the resource needs of the Superfund program and focus only on those costs that can be directly linked

Table 6-3. FY 1999 Expenditures for EPA Superfund Payroll (Millions of Dollars)

	Regions	Headquarters	Total
Charged to site accounts	52.1	1.3	**53.4**
Not charged to site accounts	155.8	66.4	**222.2**
Total	**207.9**	**67.7**	**275.6**

Note: Includes payroll associated with brownfields and federal facilities.
Source: Data provided to RFF by EPA, August 2000.

to the cleanup of sites. Excluding them from estimates of the resources required to implement Superfund is, simply put, inappropriate.

That said, our analyses of Superfund expenditure data leave a number of unresolved issues. In particular, the large amount of EPA staff time in the regional offices not charged to specific sites raises questions about where these resources are going and how much of this money is truly needed to implement the Superfund program. As shown in Table 6-3, total regional payroll costs in FY 1999 were approximately $208 million. A large proportion of these staff costs—just under $156 million, or nearly 75%—was not charged to specific sites.

In fact, how staff working in the Superfund program (both at EPA head-quarters and in the regional offices) spend their time is an interesting question that probably deserves greater scrutiny. In FY 1999 there were a total of 2,975 FTE paid for with Superfund monies (excluding the 336 FTE that worked on federal facilities and brownfields, programs not included in our study). Not surprisingly, the majority of the workforce (80%) is employed in the 10 EPA regional offices, where the Superfund program is primarily implemented.

The Superfund workforce is distributed among a number of program functions, as summarized in Table 6-4. The majority of regional staff—over 1,500 FTE, or 65%—are categorized as technical and enforcement staff. Included in this category are the remedial project managers who manage site cleanup under the remedial program and the on-scene coordinators who manage site cleanup under the removal program, as well as the attorneys and legal support personnel involved in related enforcement efforts. Almost 400 FTE in the regional offices are in functions supporting site work, such as contract and grant management, accounting, budget and financial services, and information and records management. Perhaps most surprising is the number of FTE overall devoted to accounting, contracts, and information management functions. For the Super-fund program as whole (excluding FTE for those programs not included in our study), nearly 20% of the FTE are involved in these functions.

Table 6-4. FY 1999 Distribution of EPA Superfund Workforce (Number of FTE)

	Regions	Headquarters	Total
Site technical	1,020.7	45.4	**1,066.1**
Site enforcement	501.6	107.1	**608.7**
Policy development	39.1	164.3	**203.5**
Contracts and grants	173.2	91.4	**264.6**
Information and records management	128.7	32.7	**161.4**
Accounting, budget, and financial services	87.0	74.5	**161.4**
Human resources	29.7	10.6	**40.3**
Facilities operation	27.3	5.6	**32.9**
Secretaries, clerks, and administrative	155.2	33.8	**189.0**
Management (supervisors only)	186.3	60.8	**247.1**
Total	**2,348.7**	**626.2**	**2,974.9**

Note: Totals may not add exactly because of rounding.
Source: Based on data provided to RFF by EPA, September 2000. Determined by applying ratios for FY 2000 to total number of full-time equivalents (FTE) in FY 1999. FTE associated with federal facilities and brownfields are not included.

The large amount of funds and staff time going to some program administration functions, such as contract and grant management and financial accounting systems, raises questions. This is not to say that these are not legitimate program costs—OARM and OCFO provide necessary program support—only that the data from IFMS and other information provided by EPA left us unclear as to exactly what activities were included in these categories. As a result, we have no way of determining whether all these resources are, in fact, needed to successfully implement the Superfund program.

Chapter Seven

The Future Cost of Superfund: FY 2000 through FY 2009

Congress requested that we estimate the cost to the U.S. Environmental Protection Agency (EPA) of implementing the Superfund program for the 10 fiscal years beginning in FY 2000. Although FY 2000 has ended since we began this effort, we include our estimates for that year. This chapter discusses our estimates, as well as the assumptions on which they are based. Each part of the Superfund program is addressed separately. The bulk of this chapter focuses on the cost of the remedial program, as that is where we put the lion's share of our efforts. We then provide our estimates for the entire program from FY 2000 through FY 2009, on both an annual and an aggregate basis.

Unless otherwise noted, all estimates presented in this chapter are in constant 1999 dollars. To show how our cost estimates would increase if we take into account likely inflation over the next 10 years, we also present many of our estimates in inflation-adjusted dollars.* All our baseline costs for FY 1999 reflect FY 1999 expenditures, rather than the planning or budget numbers that are typically used to describe the costs of different elements of the Superfund program.

The congressional report language requesting this study specifically indicates that we should "identify sources of uncertainty in the estimates."[1] To take into account the uncertainty associated with our cost projections, we calculate estimates under three scenarios—a base case, a low case, and a high case. The base case represents our best judgement of the future cost of the Superfund program, given the full body of our research, analyses, and interviews. The low and high cases provide a range of likely future costs, given the uncertainty about many factors in our model.

---

*See Appendix F for an explanation of how we convert past costs to constant 1999 dollars and how we adjust future dollars for inflation.

We use *average* costs and durations for each phase of the remedial cleanup process: the remedial investigation/feasibility study (RI/FS), the remedial design (RD), the remedial action (RA), and long-term response action (LTRA).* *Individual* actions, of course, will have a range of costs and require varying lengths of time to complete. This means that our projections for any specific operable unit or any specific site are likely incorrect. Nevertheless, our overall projections should accurately reflect the true range of costs, as they are part of the averages we developed.

All our estimates are based on the fundamental assumption that there will not be major changes to the Superfund program over the time period being addressed, which is consistent with the congressional report language. To accurately estimate the future costs associated with major changes to the law's liability or cleanup standards would require using dramatically different assumptions.

We discuss our cost estimates for the six primary elements of the Superfund program in the following order:

- the removal program;
- the remedial program, including
  — Fund-lead actions at current National Priorities List (NPL) sites,
  — Fund-lead actions at future NPL sites,
  — other site-specific activities (e.g., oversight, enforcement) at current and future NPL sites, and
  — efforts at non-NPL sites;
- site assessment;
- program staff, management, and support;
- program administration; and
- other programs and agencies.

Appendix G discusses in detail the methodology as well as the assumptions (e.g., cost of Fund-lead actions, durations) we use to estimate each of the six major components of the future costs of the Superfund program.

It is important to remember what is and is not included in each element. As noted in Chapter 6, we had hoped to assign program staff, management, and support costs and program administration costs to specific Superfund programs. However, the organization of EPA's expenditure data did not allow this approach. Thus—and it cannot be overemphasized—the estimates for the

---

*For the purpose of estimating future costs, we consider LTRA part of the remedial pipeline.

removal and remedial programs, in particular, do not represent their full cost. A second important point is that we project the actual expenditures that EPA will incur each year from FY 2000 through FY 2009, *not* the amount of funds that will need to be appropriated to and obligated by the Agency each year. Those charged with figuring out how much money should be appropriated and obligated each year will have to translate our estimates of expenditures into appropriations and obligations.

Cost of the Removal Program

The costs directly associated with the removal program totaled nearly $318 million in FY 1999, or 21% of that year's total Superfund expenditures. Most of the costs incurred by EPA for the removal program are extramural, specifically, the costs of paying for Fund-lead removal actions. In FY 1999 the Agency spent $220 million, or 69% of its total removal program expenditures, on Fund-lead removal actions. The balance, about $98 million, went toward other site-specific activities, such as enforcement, oversight, response support, and community and state involvement, as well as management and support costs at EPA headquarters and in the 10 EPA regions.* As noted earlier, this $98 million does not include some program staff, management, and support and program administration costs associated with the removal program.

As discussed in Chapter 2, current data are insufficient to support a meaningful prediction of the number, type, or cost of future removal actions. It is clear that the removal program is budget-driven, in that EPA regions are conducting as many emergency responses and removal actions as current budget levels allow. Arguably, Congress or EPA could, as a matter of policy, increase or decrease the budget of the removal program. Absent a thorough evaluation in all 50 states (and 9 territories) of the number and severity of releases and threatened releases of hazardous substances, it is difficult to say whether current funding is appropriate, or whether it should be more or less. Thus, we assume level funding for the removal program for FY 2000 through FY 2009, using FY 1999 expenditures of $318 million as the baseline, and estimate that total removal program costs during this period will be $3.18 billion (or $3.56 billion, adjusted for inflation).

----

*We include both EPA and Department of Justice (DOJ) enforcement costs together under site-specific enforcement.

Cost of the Remedial Program

We estimate the cost of the remedial program in five parts. First, we consider the future cost of Fund-lead actions at current NPL sites—that is, the 1,245 nonfederal final and deleted sites on the NPL at the end of FY 1999. Second, we project the cost of Fund-lead actions at sites added to the NPL from FY 2000 through FY 2009, which we refer to as future NPL sites. Third, we add the cost of Fund-lead actions at current and future NPL sites to arrive at a total estimate for Fund-lead actions. Fourth, we estimate the cost of other site-specific activities at NPL sites (current and future), including the costs associated with oversight, enforcement, response support, community and state involvement, and five-year reviews. Lastly, we address the costs of efforts undertaken at non-NPL sites. As with the estimate of the cost of the removal program, a number of costs associated with the remedial program—such as policy and management staff in EPA headquarters and rent—are not included here but are addressed under other cost elements discussed below.

Cost of Fund-Lead Actions at Current NPL Sites

We estimate future costs under three scenarios, using the same basic methodology but modifying certain assumptions to develop alternative estimates. We first describe our approach for the base case, which represents our best estimate of future costs, and then change some of the basic assumptions to develop low and high estimates. Each scenario starts from the same baseline, described below.

Our starting point is the end of FY 1999, at which time there were 1,245 nonfederal sites on the NPL: 1,055 final NPL sites and 190 deleted NPL sites. These sites have approximately 2,200 remedial operable units, or an average of 1.8 operable units per site. In this section we provide our estimates for the cost of Fund-lead actions—that is, the costs of RI/FSs, RDs, RAs, and LTRAs—at these sites for FY 2000 through FY 2009. It is important to reiterate that the costs included in this section are *only* the direct extramural costs (e.g., contractor, U.S. Army Corps of Engineers) of implementing a Fund-lead pipeline action and the EPA staff costs of managing that action. Not included are the costs of overseeing actions implemented by potentially responsible parties (PRPs), or other site-specific and program costs. Thus, these estimates should *not* be taken as the "full costs" or "fully loaded costs" of the remedial program.

To estimate future costs and project when they will be incurred, we characterize the 1,245 current NPL sites as mega or nonmega sites and as teenager or nonteenager sites.* Mega sites have expected or actual total removal and remedial action costs of $50 million or more. EPA identified 112 mega sites (25 of which were construction-complete) and 1,133 nonmega sites (623 of which were construction-complete) on the NPL at the end of FY 1999. Teenager sites are those that were proposed to the NPL before FY 1987 but were not construction-complete at the end of FY 1999. Of the 1,133 nonmega sites there were 250 teenager sites and 883 nonteenager sites. In our model, we further characterize current NPL sites by site type.

Base Case: Mega Sites

A 1994 Congressional Budget Office (CBO) study of future Superfund costs first revealed the important impact that a small number of expensive sites could have on future cleanup costs.[2] Analyses we performed early in our research corroborated CBO's work. We examined EPA expenditures in FY 1999 and found that 51% of the total dollars spent by the Agency for nonremoval cleanup actions (RI/FSs, RDs, RAs, and LTRAs) was attributable to mega sites. Thus, in that year, more than half the cost of cleanup actions was spent at fewer than 10% of NPL sites.

Since mega sites are a major driver of future costs, we collected site-specific data for each of the 112 mega sites on the NPL identified by EPA. We prepared a detailed questionnaire that was sent to remedial project managers (RPMs) in each EPA regional office.† The RPMs were asked to identify all ongoing and future RI/FSs, RDs, RAs, and LTRAs at their sites and to provide the expected starting and ending fiscal year of each ongoing and future action. The RPMs were also asked to indicate whether each action was likely to be Fund-lead or PRP-lead, and if Fund-lead, to provide their best estimate of the extramural money that would be needed for each action, by year, for FY 2000 through FY 2009. Since part of FY 2000 had already passed, RPMs were asked to include both the amount already spent in FY 2000 and what would be spent during the rest of the fiscal year.

As part of the survey, the RPMs also provided information on all ongoing and future PRP-lead actions. Tracking PRP-lead actions allows us to capture

---

* As described in detail in Appendix G, we use different cost estimates for actions depending on whether they will be conducted at mega or nonmega sites, and different average durations for nonmega sites depending on whether actions will be conducted at teenager or nonteenager sites.

† See Appendix D for a description of the information collected as part of the RFF mega site survey.

Table 7-1. Summary of Remedial Pipeline Actions at Mega Sites on Current NPL, End of FY 1999

	RI/FS		Remedial design		Remedial action		LTRA	
	Fund-lead	PRP-lead	Fund-lead	PRP-lead	Fund-lead	PRP-lead	Fund-lead	PRP-lead
Actions completed[a]	129	71	78	118	52	67	4	0
Actions under way	41	53	20	40	34	79	8	9
Actions not yet started	9	13	43	97	49	120	20	51
Total	**179**	**137**	**141**	**255**	**135**	**266**	**32**	**60**

[a]Based on RFF action-level dataset. PRP-lead LTRA is also referred to as PRP LR.

any extramural Trust Fund dollars spent as part of PRP-lead actions as well as to track the total number of actions under way in each fiscal year. This latter information is used to estimate EPA's future costs associated with PRP-lead actions, such as oversight and enforcement, as discussed later in this chapter. Table 7-1 summarizes the number of cleanup actions completed, under way, and remaining at the end of FY 1999 at the 112 mega sites on the current NPL, broken down by Fund-lead and PRP-lead.

We did not collect site-specific estimates of the staff costs associated with actions at mega sites (the primary component of intramural costs). Rather, we use the intramural unit costs we calculated for mega sites from EPA's Integrated Financial Management System (IFMS) expenditure data from FY 1996 through FY 1999. As noted in Chapter 3, we estimate that intramural unit costs for mega sites will be approximately $82,000 for an RI/FS, $50,000 for an RD, and $120,000 for an RA at each Fund-lead operable unit.*

Based on the survey and our intramural unit costs, we estimate that the total extramural and intramural costs for Fund-lead actions at mega sites on the NPL from FY 2000 through FY 2009 will be $2.65 billion. By type of action, $78.4 million (3%) is for RI/FSs, $167.9 million (6%) for RDs, $2.24 billion (85%) for RAs, and $165.2 million (6%) for LTRAs. Figure 7-1 presents the estimated annual costs for Fund-lead actions at mega sites from FY 2000 through FY 2009 for the base-case scenario.

The noticeable jump in FY 2002 and FY 2003 is due primarily to the impact of two Fund-lead RAs beginning at two mega sites, one of which has an esti-

---

*Intramural costs for LTRA are included as part of response support costs, discussed later in this chapter.

Figure 7-1. Estimated Cost of Fund-Lead Actions at Mega Sites on Current NPL: Base Case, FY 2000–FY 2009 (1999$)

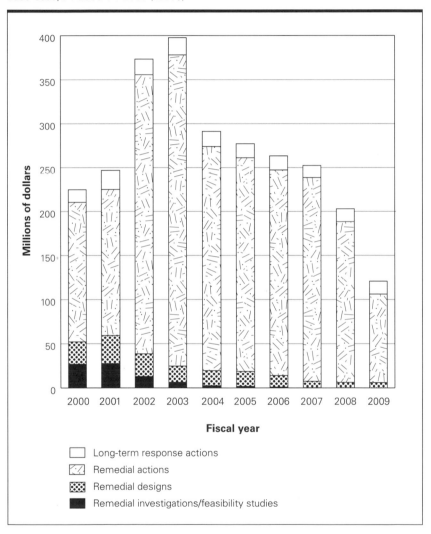

mated cost of $150 million and the other an estimated cost of $80 million in these two years alone. This is illustrative of the dramatic impact one or two mega sites can have on the total cost of the Superfund program.

Base Case: Nonmega Sites

We did not collect site-specific information on the likely future costs of conducting Fund-lead actions at the 1,133 nonmega sites on the current NPL. Instead, we determine the cleanup status of each site as of the end of FY 1999 and then make assumptions about future leads, unit costs, and durations based on our analyses of these factors.*

The initial step in estimating the number and cost of remaining cleanup actions at nonmega sites is to determine the status—RI/FS, RD, RA, or LTRA—of each operable unit at each site at the end of FY 1999. This has been done for both the 623 construction-complete nonmega sites and the 510 nonmega sites that were not construction-complete. Although most cleanup costs have already been incurred at construction-complete sites, there may be additional costs associated with finishing an RA, for example, or carrying out an LTRA.

One can measure the progress of cleanup at a site in different ways. EPA typically characterizes a site's progress based on its most advanced operable unit—in other words, the stage of the pipeline of the operable unit at which the most progress has been made. One can also judge the status of cleanup at a site based on the *least* advanced operable unit, as the U.S. General Accounting Office did in a recent analysis.[3] Both approaches, however, base a site's cleanup progress (or lack thereof) on a single operable unit. A more comprehensive approach—one necessary to fully capture the remaining costs of cleanup at a site—is to assess the progress of cleanup at *each* operable unit. This is the approach we have taken.

For nonmega sites, we identify the status of each operable unit as of the end of FY 1999. After assigning each operable unit to a phase in the remedial pipeline based on its "last action" under way or completed at the end of FY 1999, we then assume that each operable unit will move forward through the rest of the remedial pipeline. In reality, not all operable units follow this course. For instance, the record of decision (ROD) may indicate that no further action (NFA) at an operable unit is necessary, in which case there will be no subsequent RD or RA. Or the RD may result in the decision to split one operable unit into two to implement cleanup remedies more effectively, a situation we

---

*Chapter 3 and Appendix F provide information on what we refer to as the building blocks of our model, and Appendix G provides a detailed description of the RFF model.

Table 7-2. Summary of Remedial Pipeline Actions at Nonmega Sites on Current NPL, End of FY 1999

	RI/FS		Remedial design		Remedial action		LTRA	
	Fund-lead	PRP-lead	Fund-lead	PRP-lead	Fund-lead	PRP-lead	Fund-lead	PRP-lead
Actions completed	610	605	321	690	208	492	3	10
Actions under way	108	107	63	126	118	265	48	97
Actions not yet started	36	37	137	236	155	399	123	253
Total	**754**	**749**	**521**	**1,052**	**481**	**1,156**	**174**	**360**

Note: PRP-lead LTRA is also referred to as PRP LR.

refer to as OU growth. Since it is not possible to predict which operable units will be designated NFA or which will be subdivided, we address these factors by adjusting our unit costs, as appropriate.

Table 7-2 shows the number of remedial pipeline actions completed and under way for the 1,133 nonmega sites on the NPL at the end of FY 1999, as well as the estimated number of future actions at these sites. Using these data as our starting point, we project the start and completion dates of subsequent pipeline actions. We determine the lead of future actions based on EPA planning data as well as by using historical information on the leads of different actions by site type.

For nonmega sites, we vary the extramural unit costs for each stage of the remedial pipeline (RI/FS, RD, and RA) vary depending on the type of site (e.g., chemical manufacturing or mining). We use the same average cost, however, for intramural costs for RI/FSs, RDs, and RAs. As noted in Chapter 4, we assume that 45% of all Fund-lead RAs will be followed by LTRAs, and that LTRAs will be necessary for nine years at an annual cost of $400,000.* Of course, in some cases LTRA activities will be needed for less than nine years, and in others for longer.† EPA provided RFF with information on the expected completion date of all ongoing actions, as well as information on the start date and lead of the next planned action at each operable unit. We use this informa-

---

*We also assume that 45% of PRP-lead RAs will be followed by groundwater or surface water restoration activities, referred to as PRP LR.

†It is only in cases where LTRA activities are needed for more than 10 years that the states will pick up operational and financial responsibility for these efforts.

tion, where available, to determine when future actions will be conducted. For those actions for which we do not have planning data, we base the timing on our average durations of remedial pipeline actions. Different durations are used for actions conducted at teenager and nonteenager sites.

Given these assumptions, we estimate that the total cost (extramural and intramural) of Fund-lead actions at nonmega sites currently on the NPL from FY 2000 to FY 2009 will be approximately $1.78 billion. As was the case for mega sites, RAs account for most of the costs. With respect to each phase of the remedial pipeline, we estimate that $80.1 million will be needed for RI/FSs (5%), $135.3 million for RDs (8%), $1.17 billion for RAs (66%), and $388 million for LTRAs (22%).* Figure 7-2 shows the total estimated costs for Fund-lead actions at nonmega sites on an annual basis from FY 2000 through FY 2009. From FY 2000 through FY 2005, annual costs fluctuate between approximately $210 million and $237 million, with a peak in FY 2004. In FY 2006 the total cost of Fund-lead actions at nonmega sites on the current NPL begins to decline steadily and is only about $68 million in FY 2009. Some additional costs to complete cleanup at these sites will be incurred after FY 2009.

Base Case: Total

The sum of the mega site and nonmega site estimates represents the total expected cost of Fund-lead actions (extramural and intramural) in the base case from FY 2000 through FY 2009 for sites on the NPL at the end of FY 1999. The total projected cost of Fund-lead RI/FSs, RDs, RAs, and LTRAs from FY 2000 through FY 2009 is about $4.43 billion (or $4.89 billion, adjusted for inflation). Of this total, about 60% will be needed for actions at mega sites, which constitute fewer than 10% of the total number of sites.

Figure 7-3 shows the total estimated cost of Fund-lead actions in the base case on an annual basis from FY 2000 through FY 2009, broken out by mega sites and nonmega sites. For FY 2000 and FY 2001, we estimate that the total cost of these actions will be about $460 million, which is about $18 million less than what EPA spent on these actions in FY 1999. From FY 2002 through FY 2005, our estimates of EPA expenditures are considerably higher than the amount spent by EPA in FY 1999, peaking in FY 2002 and FY 2003 because of the impact of Fund-lead RAs at two mega sites, as previously noted. Beginning with FY 2006, these costs fall below their FY 1999 level and begin to decline steadily to about $188 million in FY 2009. Figure 7-3 also shows the results when adjusted for inflation.

---

*Percentages do not add to 100% because of rounding.

Figure 7-2. Estimated Cost of Fund-Lead Actions at Nonmega Sites on Current NPL: Base Case, FY 2000–FY 2009 (1999$)

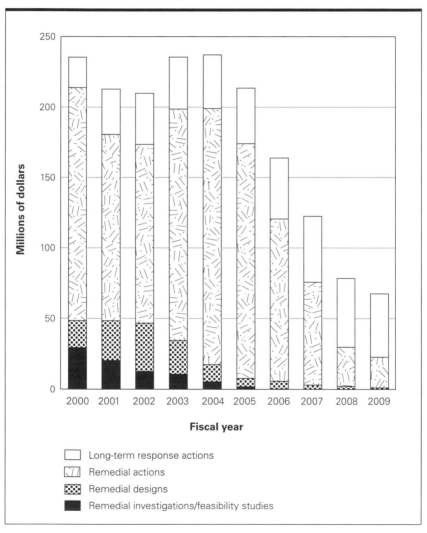

Figure 7-3. Estimated Cost of Fund-Lead Actions at Current NPL Sites: Base Case, FY 2000–FY 2009 (1999$ and Inflation-Adjusted Dollars)

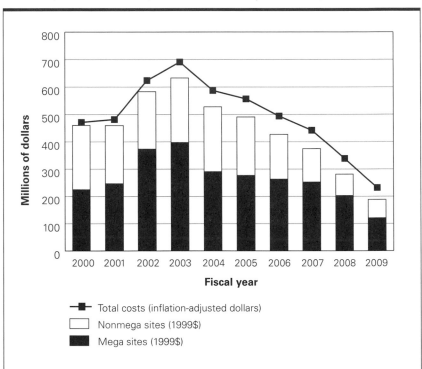

Of the $4.43 billion dollars in projected costs of Fund-lead actions for sites on the current NPL from FY 2000 through FY 2009, we estimate that nearly $553 million will be required for LTRAs, or roughly 12% of the total. A key issue regarding the future of post-construction costs at NPL sites is the ability of states to take over LTRAs at groundwater or surface water operable units if restoration goals have not been met after 10 years. States have thus far shown the ability to finance operation and maintenance activities. These activities, however, have generally not included the expensive pump-and-treat systems needed for groundwater restoration, which is the primary distinction between ordinary operation and maintenance and LTRA. We project that the total number of LTRA completions from FY 2000 through FY 2009 will be relatively few. During this period, we estimate that 90 operable units with LTRAs will shift from EPA to state implementation—83 at nonmega sites and 7 at mega sites. Assuming that LTRA is not completed at any of these sites before responsibility is transferred to the states, the total LTRA costs to the states for taking over these LTRA costs as part of their operation and maintenance obligations will be approximately $167.2 million from FY 2000 through FY 2009. The financial burden on the states will be more pronounced after FY 2009, when the number of LTRA completions will rise. Since operation and maintenance activities associated with groundwater restoration can last decades, the total expense of maintaining a pump-and-treat system at a mega site could, for example, cost between $40 million and $100 million ($2 million to $5 million per year for 20 years).

Low Case

EPA has recently stated that the cost of cleanup has been reduced by approximately 20% because of the administrative reforms to the remedy selection process implemented by the Agency in the 1990s.[4] As noted in Chapter 3, EPA attributes these cost savings to the use of presumptive remedies, the determination of reasonably anticipated future land uses to allow cleanups to be tailored to specific sites, and the phased approach to defining objectives and methods for groundwater cleanups. These changes began to affect remedy decisions made in FY 1997; thus RAs implemented in FY 2000 would be expected to cost less.

To reflect the potential cost savings, our low-case scenario incorporates a reduction in unit costs for future RAs (extramural costs only). Keeping other components of the model unchanged, we discount all future RAs at mega sites and nonmega sites by 20%. With these cost savings, we estimate that the overall cost of Fund-lead RAs for sites on the current NPL from FY 2000 through FY 2009 will be approximately $2.95 billion, almost $460 million less than in the base case. Total projected cost in the low case from FY 2000 through FY 2009 is about $3.97 billion (or $4.38 billion, adjusted for inflation), or a little more than 10% less than in the base case.

As mentioned in Chapter 3, however, our analysis of RA expenditures over time does not show the cost savings that EPA has predicted. In fact, when comparing RAs completed between FY 1996 through FY 1999 with those completed between FY 1992 through FY 1995, we detect an *increase* in the average cost of an RA. Since EPA based its projections of cost savings on estimates in recently signed RODs, it may be that the 20% reduction in the cost of cleanups the Agency is expecting has yet to materialize.

High Case

We also estimate a high-case scenario to examine the potential impact on the need for EPA funds if cleanups at sites on the current NPL are more costly than estimated in the base case. In the high case we make two changes to our base-case estimates of the future cost of Fund-lead actions at mega sites. First, we substitute RFF's estimated unit costs of Fund-lead actions (RI/FSs, RDs, and RAs), which are generally higher than the average of the costs estimated by the remedial project managers (RPMs) in the information they provided to RFF.* We do not substitute our estimated LTRA costs, however, but use those estimated by the RPMs. We also include in the high-case scenario the cost of possible additional actions identified by the RPMs. For a number of mega sites, RPMs indicated that there might be a need for further remediation at their sites but said it was too early to predict how many actions and how much they would cost. They identified 70 to 100 such potential additional actions. Most were expected to be PRP-lead, so the actual impact in terms of Trust Fund dollars would be relatively small and would be primarily for additional oversight and response support. About 20% of these additional actions, however, were potentially Fund-lead. In total, there are 18 such actions—4 RI/FSs, 5 RDs, 6 RAs, and 3 LTRAs—for which we use RFF average extramural and intramural unit costs,† spreading the dollars evenly across FY 2005 through FY 2009.

With these assumptions, from FY 2000 through FY 2009, the total cost of Fund-lead actions at current NPL sites will be approximately $5.08 billion (or $5.62 billion, adjusted for inflation), which represents an increase of about 17% ($655 million) over the base case. More than 60% of the increase is attributable to using our extramural unit costs instead of the estimates provided by the RPMs. Figure 7-4 shows the annual estimates under all three scenarios for the cost of Fund-lead actions at current NPL sites from FY 2000 through FY 2009.‡

Cleanup Progress at Current NPL Sites

By the end of FY 2009, we estimate that 1,198 (96%) of the 1,245 nonfederal NPL sites that were final or deleted at the end of FY 1999 will be construction-

---

*We do not substitute our unit costs for actions that were ongoing at the beginning of FY 2000 or actions that started during FY 2000.

†For LTRA at mega sites, we assume an annual cost of $2 million.

‡See Table H-1 in Appendix H for the annual cost estimates for Fund-lead actions at current NPL sites under the three scenarios from FY 2000 through FY 2009.

Figure 7-4. Estimated Cost of Fund-Lead Actions at Current NPL Sites:
Three Scenarios, FY 2000–FY 2009 (1999$)

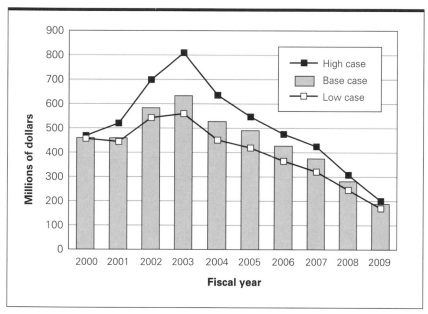

complete.* Our projection of when sites will reach construction completion differs from EPA's. In its most recent strategic plan, the Agency stated that it expects to complete construction at 1,105 NPL sites by the end of FY 2005,[5] whereas our model indicates that only 994 will be construction-complete. Some of the difference may be that EPA's projections also account for progress at federal facilities, which are not included in our analysis. However, EPA has itself noted that beginning in FY 2002, it does not expect to be able to maintain the recent pace of cleanup.[6]

Table 7-3 indicates the number of RI/FSs, RDs, RAs, and LTRAs—both Fund-lead and PRP-lead—that we project will be under way or will not yet have begun at current NPL sites at the end of FY 2009.† Most remedial pipe-

---

*See Appendix G for a description of the methodology we use to project when sites will reach construction completion and Table H-2 in Appendix H for our projections of the number of sites on the current NPL that will be construction-complete each year.

†Table H-3 in Appendix H shows the estimated number of actions that will not yet be completed at the end of each fiscal year from FY 2000 through FY 2009. Table H-4 in Appendix H presents these same data on a percentage basis.

Table 7-3. Estimated Number of Remedial Pipeline Actions Not Yet Completed at Current NPL Sites, End of FY 2009

	Mega sites		Nonmega sites	
	Fund-lead	PRP-lead	Fund-lead	PRP-lead
Under way				
RI/FS	0	0	0	0
Remedial design	1	0	1	7
Remedial action	10	25	21	77
LTRA and PRP LR	19	52	83	269
Subtotal	**30**	**77**	**105**	**353**
Not yet started				
RI/FS	0	0	0	0
Remedial design	0	0	0	2
Remedial action	0	0	1	8
LTRA and PRP LR	2	4	5	26
Subtotal	**2**	**4**	**6**	**36**

line actions at current NPL sites, with the exception of LTRAs, should be completed by the end of that year.

Cost of Fund-Lead Actions at Future NPL Sites

The most controversial aspect of this study is almost certainly the estimate of the cost of future NPL sites—the sites added to the NPL from FY 2000 through FY 2009. Undoubtedly, this is the most difficult number to predict with any confidence. As discussed in Chapter 5, three key factors affect the cleanup cost of future NPL sites: the number of nonfederal sites added to the NPL each year, the number of mega sites added to the NPL each year, and the proportion of remedial pipeline actions that are Fund-lead.

Our three scenarios include variations in these three factors. The estimates under each represent the future cost of Fund-lead actions—RI/FSs, RDs, RAs, and LTRAs at sites added to the NPL from FY 2000 through FY 2009 and their associated intramural costs. (LTRA costs for future NPL sites are minimal in our time frame, as EPA will not begin to incur these costs until FY 2009, when a few operable units reach the post-construction phase of the remedial pipe-line.) We address other site-specific costs associated with future NPL sites (e.g., oversight and enforcement) as well as general program staff, management, and support and program administration costs later in this chapter. Thus, the estimates presented in this section address *only* the extramural and intramural costs of Fund-lead remedial pipeline actions at sites added to the NPL from FY 2000 through FY 2009.

The logic behind our assumptions for the three scenarios was explained in Chapter 5. Each scenario is described briefly below, followed by our estimates of the cost (extramural and intramural) of Fund-lead actions from FY 2000 through FY 2009.

Base Case

Our base case is the scenario that we believe to be most likely, given the results of our research and interviews. For FY 2000 we use, in our model, the actual number of final sites added to the NPL, 36.* For each year from FY 2001 through FY 2009, we assume that 35 sites will be added to the NPL.† For each year, we also assume that two of the newly listed sites are mega sites, including in FY 2000 (when, in fact, two mega sites—Leviathan Mine in California and Midnite Mine in Washington State—were listed). We also assume that the proportion of Fund-lead actions at mega sites will reflect recent history (FY 1996 through FY 1999); that is, that 80% of RI/FSs, 35% of RDs, and 25% of RAs will be Fund-lead. However, for nonmega sites we assume that the percentage of Fund-lead actions will be slightly higher than in recent years; that is, that 75% of RI/FSs, 40% of RDs, and 40% of RAs will be Fund-lead. We make this adjustment because EPA expects a higher proportion of Fund-lead actions at nonmega sites;[7] moreover, absent major changes in how states allocate their resources, even an "average" NPL site is beyond the current funding capabilities of most state superfund programs.

To arrive at our cost estimates, we use the same basic methodology we applied to current NPL sites, with a few variations. We use our extramural and intramural unit costs for both mega sites (in place of site-specific data) and nonmega sites for each of the major cleanup actions: RI/FSs, RDs, RAs, and LTRAs. Since it is not possible to forecast the breakdown of site types on an

---

*Technically, 37 sites were added to the NPL in FY 2000, but one, Georgia-Pacific Corp. in North Carolina (Region 4), was later withdrawn.

†The number of sites added to the NPL each year in each case represents the number of sites that will be listed as *final* NPL sites in that year.

annual basis, we use the average costs over all site types, for both mega sites and nonmega sites. To splay the costs of these actions from FY 2000 through FY 2009, we use two sets of average durations, one for actions that will be conducted at mega sites and another for actions that will be conducted at non-mega sites.* And as with current NPL sites, we adjust the costs of different phases of the pipeline to take into account OU growth and RODs calling for no further action.

Based on analysis of recent Comprehensive Environmental Response, Compensation, and Liability Information System (CERCLIS) data, we assume that each mega site has 3.8 operable units, and each nonmega site, 1.6 operable units.[8] Lastly, we delay by one year the onset of actions at successive operable units at individual sites to take into account "OU lag," the average time between the start of cleanup at successive operable units at a single site.

In the base case, we estimate that the total cost to EPA of implementing Fund-lead cleanup actions at sites added to the NPL from FY 2000 through FY 2009 will be approximately $854.6 million (or $996 billion, adjusted for inflation). Of this total, about $678.3 million will be needed for Fund-lead actions at nonmega sites and $176.3 million for similar actions at mega sites.

Low Case

The low case assumes that future listings will be consistent with recent trends. We use the actual number of final NPL sites, 36, in FY 2000. In each year from FY 2001 through FY 2009, we assume that 23 sites will be added to the NPL, the average number of sites listed over the past five years (FY 1996 through FY 2000). In our low case, we assume that each year one of the newly listed sites is a mega site, with the exception of FY 2000, when two mega sites were listed. As with the total number of sites, the annual addition of one mega site is consistent with the trend over the past five years.

We assume that for mega sites, 80% of RI/FSs, 35% of RDs, and 25% of RAs will be conducted by EPA and paid for with Trust Fund dollars. At non-mega sites, we assume that 75% of RI/FSs, 30% of RDs, and 30% of RAs will be Fund-lead. These percentages reflect the overall proportion of Fund-lead actions from FY 1996 through FY 1999. As with the low-case scenario for sites on the current NPL, we assume that RAs conducted at future NPL sites will cost 20% less than our estimated unit costs (extramural only).

---

*See Table G-10 in Appendix G for the average durations we use for actions that will be conducted at future mega and nonmega sites.

In the low case, we estimate that the total cost to EPA of Fund-lead actions at future NPL sites from FY 2000 through FY 2009 will be approximately $519.5 million (or $601.6 million, adjusted for inflation). Of this total, about $412.1 million will be needed for Fund-lead actions at nonmega sites and $107.4 million for Fund-lead actions at mega sites.

High Case

In this scenario, we assume that in each year from FY 2001 through FY 2009, 49 sites will be added to the NPL. Again, in FY 2000, we use the actual number, 36 sites. Of the 49 annual additions to the NPL, we assume that 3 will be mega sites, except in FY 2000, when 2 mega sites were added. The annual addition of 49 NPL sites is based on our interviews with officials in the 10 EPA regions and corresponds to their low estimate of the number of sites likely to be added to the NPL. Further, we assume that there will be an increase in the proportion of Fund-lead actions at both mega sites and nonmega sites. This takes into account both the information we gathered in our interviews with the EPA regions and nine states and the possibility that PRPs will not finance as much of the work at mega sites as they have in recent years. For mega sites, we assume that 80% of RI/FSs, 35% of RDs, and 35% of RAs will be Fund-lead; for nonmega sites, we assume that 75% of RI/FSs, 40% of RDs, and 40% of RAs will be Fund-lead.

In the high case, we estimate that the cost incurred by EPA for Fund-lead actions at these sites from FY 2000 through FY 2009 will be approximately $1.14 billion (or $1.33 billion, adjusted for inflation). Of this total, about $878.1 million will be needed for Fund-lead actions at nonmega sites and $257.7 million for Fund-lead actions at mega sites.

Figure 7-5 compares the annual estimates for Fund-lead cleanup actions at future NPL sites under the three scenarios.* From FY 2000 through FY 2004, the differences among the cost estimates for the three scenarios are relatively small. For instance, even though we estimate that the high-case cost will be about double the low case in FY 2004, the disparity that year is estimated to be only $37 million. However, over the next five years, the annual difference between the two scenarios becomes more pronounced. By FY 2009, the difference between the low and high cases in estimated costs of Fund-lead actions at future NPL sites is more than $190 million. The spread becomes even larger

---

*See Table H-5 in Appendix H for the annual cost estimates for Fund-lead actions at future NPL sites under each scenario from FY 2000 through FY 2009.

Figure 7-5. Estimated Cost of Fund-Lead Actions at Future NPL Sites: Three Scenarios, FY 2000–FY 2009 (1999$)

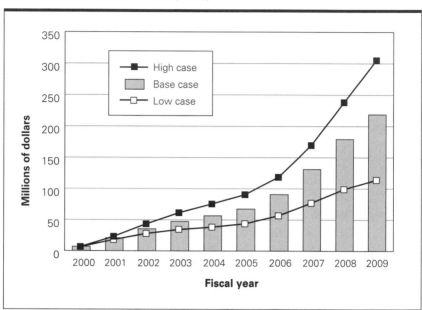

after FY 2009, when more of the sites listed in FY 2000 and after will have operable units in the RA phase of the remedial pipeline.

To better understand the impact of the different scenarios, it is worth extending the time line and looking at the total cost of Fund-lead actions through FY 2030, shown in Figure 7-6. The total cost for the low case is about $1.51 billion; for the base case, about $3.02 billion; and for the high case, about $4.61 billion. Thus, even though the cost implications of the composition of the future NPL are relatively small from FY 2000 through FY 2009, decisions made during this period will have a dramatic impact on the long-term cost of the Superfund program. (Costs begin to drop around FY 2014 because, in this set of estimates, we assume that no sites are added to the NPL after FY 2009. The cost curves would continue at a higher level if, in the model, we continued to add new NPL sites in FY 2010 and beyond.)

As emphasized in Chapter 5, there is enormous uncertainty regarding the composition and cost of the sites added to the NPL in the future. This is because which sites are added to the NPL is, fundamentally, a policy decision—one made by the Administrator of EPA, in theory, and in fact by the Assistant Administrator of the Office of Solid Waste and Emergency Response

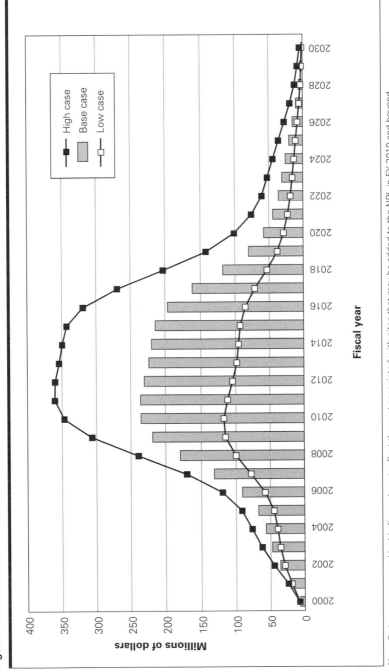

Figure 7-6. Estimated Cost of Fund-Lead Actions at Future NPL Sites: Three Scenarios, FY 2000–FY 2030 (1999$)

Note: Estimates presented in this figure do not reflect the cost associated with sites that may be added to the NPL in FY 2010 and beyond.

(OSWER), in concert with EPA's 10 regional offices and the states. Additionally, fluctuations in the national or state economies could greatly affect the number and types of sites states want added to the NPL. With these caveats, however, barring a major policy change regarding what type of sites should be listed, and taking into account what we learned in our interviews and discussions with state officials and EPA regional and headquarters staff, the base case reflects the most likely picture of the future NPL. Our low and high cases illustrate the likely range of future costs.

Total Cost of Fund-Lead Actions at NPL Sites

Before we proceed to estimate the other two categories of remedial program costs, we sum the estimates for direct cleanup activity at current and future NPL sites to obtain an estimate of total likely cost of Fund-lead actions at NPL sites from FY 2000 through FY 2009. Since the base cases represent the best estimates of likely future costs for both the current and the future NPL, adding these estimates provides the best overall estimate of the future cost to EPA of carrying out Fund-lead actions at NPL sites. From this point forward, we refer to this sum as the base case. Combining the two low estimates and the two high estimates provides the best estimate of the likely range of future costs. From this point forward, we refer to these as the low case and the high case, respectively.

In total, we estimate that the cost incurred by EPA in our base case for Fund-lead actions at NPL sites from FY 2000 through FY 2009 will be approximately $5.28 billion (or $5.89 billion, adjusted for inflation). Our estimates for the low and high cases are $4.49 billion (or $4.98 billion, adjusted for inflation) and $6.22 billion (or $6.94 billion, adjusted for inflation), respectively. Thus, the lower estimate is about 15% less than the base case, and the high case is about 18% more than the base case. Figure 7-7 shows the estimates for each scenario and, for the purpose of comparison, indicates the equivalent level of spending for Fund-lead actions at NPL sites in FY 1999.

Cost of Other Site-Specific Activities at NPL Sites

In addition to the direct costs of Fund-lead actions, there are additional site-specific activities associated with cleaning up sites on the NPL. For example, EPA oversees remedial pipeline actions implemented by potentially responsible parties (PRPs) (who perform the majority of site cleanups), and both EPA and the U.S. Department of Justice (DOJ) conduct PRP searches, negotiate

Figure 7-7. Estimated Cost of Fund-Lead Actions at Current and Future NPL Sites: Three Scenarios, FY 2000–FY 2009 (1999$)

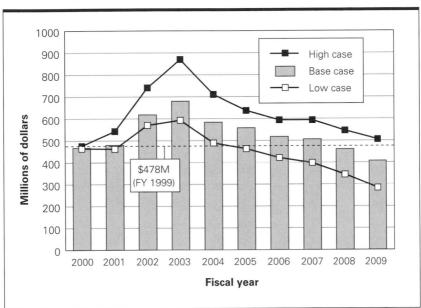

Note: FY 1999 expenditures for these types of costs were $478 million.

with PRPs, and recover funds from PRPs who do not implement cleanups themselves but are liable for the costs under the Comprehensive Environmental Response, Compensation, and Liability Act (CERCLA). Other site-specific activities include response support (such as laboratory analysis), community and state involvement (such as providing technical assistance grants), and five-year reviews, all of which are implemented at both Fund-lead and PRP-lead sites. These other site-specific activities are integral to the Superfund program.

Ideally, it would be possible to calculate an average cost for each of these site-related activities at the action, operable unit, or site level. However, with the exception of five-year reviews, expenditure data in IFMS are not coded in a way that facilitates this type of analysis.

As an alternative, we tie the annual costs of these activities for FY 2000 through FY 2009 to the overall level of work in the remedial program. To do this, we developed five scaling factors, which represent different measurements of the annual workload of the Superfund program, such as the total number of PRP-lead remedial pipeline actions. The use of these scaling factors provides a transparent, albeit imperfect, method of estimating the costs of these other site-specific activities in relation to the total workload of the remedial program.

For each year from FY 2000 through FY 2009, we calculate each scaling factor as a percentage of its measured value in FY 1999. We then adjust the FY 1999 expenditures up or down by the appropriate scaling factor percentages to obtain expected annual expenditures in future years. Measurements of the Superfund workload will, of course, depend on which of the three scenarios—base, low, high—is used. Thus, these other site costs also vary under each scenario.

Oversight of PRP-Lead Actions

For every remedial pipeline action conducted by a PRP at an NPL site, whether it is an RI/FS, RD, RA, or LTRA (referred to as PRP LR when conducted by PRPs), there are EPA staff who ensure that Superfund policies and regulations are followed. (In practice, these oversight activities, while funded by EPA, are often carried out by the U.S. Army Corps of Engineers, private contractors, or states, as well as by EPA regional staff.) To estimate the annual oversight dollars the Agency will need from FY 2000 through FY 2009 under the three scenarios, we use an adjusted PRP action scaling factor, which is the total number of PRP-lead actions (RI/FSs, RDs, RAs, and PRP LRs) under way at the end of each fiscal year. We know from our analyses that more effort is required to oversee mega sites. To account for this, we multiply oversight costs for each action at mega sites by two, because we assume the work is twice that of a nonmega site.* Oversight of PRP LR, however, requires less work than oversight of the other PRP-lead actions, so we multiply each PRP LR by 20%.†

In FY 1999, $33.2 million was spent on oversight of PRP-lead remedial pipeline actions at NPL sites (not including PRP-lead removal actions). We adjust future oversight expenses in relation to this amount. For example, at the end of FY 2004 in our base case, the total number of adjusted PRP-lead remedial pipeline actions under way is 29% less than the total number at the end of FY 1999. Thus, we adjust the oversight dollars EPA will need in FY 2004 downward by 29% from FY 1999 levels, to $23.7 million.

---

*We use a multiplying factor of two because it is approximately the same ratio as intramural costs for Fund-lead actions at a mega site compared with similar costs at a nonmega site.

†The decision to use a factor of 0.2 for PRP LR is based on discussions with EPA. Although it is clear that oversight of PRP LR requires less work, there are no data to quantify the actual value of this factor.

In the base case, we estimate that from FY 2000 through FY 2009 EPA will need approximately $230.2 million to oversee PRP-lead actions at NPL sites. In comparison, we estimate that oversight costs in the low case and high case will be $221.1 million and $243.5 million, respectively. Over this period, we estimate that the level of oversight costs will decline each year in the low and base cases. We project a similar overall trend for the high case, although there are small increases in FY 2008 and FY 2009.

Enforcement

Site-specific enforcement activities can be separated into two general categories: cost recovery and efforts to "maximize PRP involvement," which includes PRP searches and settlement negotiations. Both DOJ and EPA carry out enforcement activities in these two categories. According to DOJ, about 80% of its enforcement effort relating to NPL sites goes to preparation and litigation of cost recovery cases, whereas only about 25% of EPA's enforcement effort is related to cost recovery. The reverse is true for maximizing PRP involvement. EPA says that about 75% of its enforcement effort goes to the earlier stages of the process, such as finding PRPs and negotiating settlement agreements. In contrast, only 20% of DOJ's effort is associated with these activities. Thus, in terms of FY 1999 expenditures, DOJ spent approximately $20.2 million on cost recovery activities for NPL sites and $5.1 million on other NPL site-related enforcement activities. In comparison, EPA spent $22.9 million on efforts to maximize PRP involvement and $7.6 million on cost recovery. We use these FY 1999 expenditure levels as the basis for adjusting the future costs of these enforcement activities.

We use different scaling factors to estimate the future costs associated with cost recovery and efforts to maximize PRP involvement. For the cost recovery portion, we use two scaling factors—one for DOJ and one for EPA. Cost recovery activities generally are not started in earnest until site cleanup is well under way. According to DOJ officials, in general, expenditures for its cost recovery efforts begin three years after the initiation of the last RA at a site, and the costs are typically spread over five years, the average time DOJ needs to pursue a cost recovery case.* Officials from EPA's Office of Enforcement and Compliance Assurance (OECA) say that their efforts typically begin two years after the start of the final RA at a site. By starting a year earlier, the Agency has

---

*It should be noted that some cases take much longer to conclude, and some, much less. Five years represents the likely average, according to information provided to RFF by DOJ officials, January 2001.

a head start to develop the cost recovery litigation referral package that is sent to DOJ (which includes the complaint and documentation of the costs incurred and the work performed at the site).[9] As for DOJ, we assume that EPA's cost recovery efforts are also spread over five years.

To calculate the DOJ and EPA cost recovery scaling factors, we total the number of sites for which cost recovery efforts are likely to still be under way. This allows us to project the agencies' cost recovery workload. We then compare these numbers with those from FY 1999 and adjust our cost projections up or down accordingly. For example, to determine DOJ's workload in FY 2000, we sum the number of NPL sites that had their last RA start from FY 1993 through FY 1997 (this takes into account both the three-year lag before DOJ begins its cost recovery efforts and the five-year period in which it generally conducts such efforts).

For all other enforcement costs, we use an NPL activity scaling factor, which serves as a rough estimate of the total number of "active" NPL sites—those that are not yet construction-complete. The NPL activity scaling factor is calculated by taking the total number of final and deleted nonfederal NPL sites at the end of FY 1999 (1,245), subtracting the total number of these sites that are projected to be construction-complete at the end of the year, and adding the total number of new nonfederal sites added to the NPL since the end of FY 1999. For example, for FY 2000, the NPL activity is $(1{,}245 - 734 + 36) = 547$. In the base case for FY 2009, this number is 364, suggesting that overall NPL activity will decrease by nearly 40% from FY 1999.

Applying the scaling factors to site-specific enforcement costs from FY 2000 through FY 2009 under the three scenarios results in estimates of $501.4 million for enforcement costs in the base case, $476.1 million for the low case, and $530.9 million for the high case. Thus, the range of average annual costs is $47.6 million to $53.1 million, compared with the $55.8 million spent by the Agency in FY 1999.[10]

Response Support

Response support is a catchall category that includes management and support functions directly related to site cleanup (such as laboratory analysis, contract management, and health assessment support). To estimate the future cost to EPA of carrying out site-specific response support efforts, we use an adjusted all-actions scaling factor, similar to the adjusted PRP action scaling factor used to estimate oversight expenses. The difference is that it includes *all* actions at NPL sites (both Fund-lead and PRP-lead) to create an annual workload factor. Again, each action at a mega site is multiplied by two to reflect the Agency's increased effort at mega sites, and LTRAs and PRP LRs are multiplied by 20% to reflect the lesser effort these actions require. We again use FY 1999 as the base year and calculate the scaling factor for each scenario. Since we predict that the overall number of actions will decrease each year as cleanup work is completed (from FY 2000 through FY 2009, the number of actions completed each year at current NPL sites will outpace the number of actions generated by future NPL sites), response support costs decline in each scenario.

Using the adjusted all-actions scaling factor, we estimate that the total response support costs to EPA from FY 2000 through FY 2009 will be about $115.8 million in the base case. In the low case, we estimate response support will cost about $105.8 million, and in the high case, about $127.2 million. Thus, the range of average annual costs is $10.6 million to $12.7 million, compared with the approximately $15.3 million the Agency spent in FY 1999.

Community and State Involvement

We also use the adjusted all-actions scaling factor to estimate EPA's expected costs for state and community involvement activities at current and future NPL sites. As for response support, we project that EPA's expenditures for community and state involvement will decrease from FY 2000 through FY 2009 as the total number of actions declines. In the base case, we estimate that the Agency will need about $75 million for state and community involvement at NPL sites from FY 2000 through FY 2009. In the low case, we estimate that EPA will need about $68.4 million, and in the high case, $82.3 million. Thus, the range of average annual costs is $6.8 million to $8.2 million, compared with the $9.9 million EPA spent in FY 1999 for these same activities.

Five-Year Reviews

Five-year reviews are part of EPA's oversight efforts and are conducted at all NPL sites, that is, at sites with both Fund-lead and PRP-lead actions.[11] Unlike the other site-specific support activities discussed in this section, we estimate the cost of five-year reviews on a site-by-site basis. To do this, we first determine when NPL sites will need reviews from FY 2000 through FY 2009, based on the requirements specified in CERCLA and current EPA policy.[12] For a number of NPL sites, EPA provided us with the dates when reviews would be due. For the balance of sites, we estimate five-year review due dates based on the start date of the first RA at the site (the trigger for statutory reviews) or the construction-complete date of the site (the trigger for policy reviews).[13]

As discussed in Chapter 4, we assume that the average cost of conducting a five-year review is $25,000 at a nonmega site and $50,000 at a mega site. Using these averages and the assumptions noted above regarding when sites on the NPL will require five-year reviews, we estimate that five-year reviews from FY 2000 through FY 2009 will cost EPA about $52.6 million in the base case and—because there is no variation in the timing of actions at NPL sites among the three scenarios—in the low and high cases as well. Our estimate is for sites on the current NPL, because based on the average durations we use, none of the future sites will, by FY 2009, have reached the point when their first five-year reviews will be due.

On an annual basis, the Agency will need to spend about $5.3 million to conduct five-year reviews for NPL sites, which is more than 10 times what the Agency appears to have spent in FY 1999, according to IFMS expenditure data. There are two reasons for this large difference in cost. First, we assume that EPA will perform all the five-year reviews when they are due, which historically has not been the case, plus clear the existing backlog. In FY 1999, EPA completed 113 five-year reviews at nonfederal NPL sites, but we estimate that the Agency will have to conduct about twice that many each year. Second, total expenditures in FY 1999, as captured in IFMS, likely underrepresent the money actually spent on five-year reviews in that year.

Several state officials said in interviews that our five-year review cost estimates were conservative, and they provided anecdotal information suggesting the true cost might be considerably higher. However, the impact on the total Superfund program cost would still be small. For example, if the cost of five-year reviews were in fact twice that used in our model, total costs would be $105.1 million from FY 2000 through FY 2009. On an annual basis, this would represent an increase of just over $5 million, or less than 0.5% of total annual Superfund expenditures in FY 1999.

It should be noted that our estimates do not include the cost of possible "remedy failures." At some percentage of sites, five-year reviews or other monitoring efforts will likely uncover the need for additional capital investment in existing remedies. At some sites, a new remedy may be required. And, it is almost certain that some remedies will fail, which means that, in some cases, a future investment of Trust Fund monies will be needed, although there is insufficient data to estimate how often this will happen, and how much it will cost.

Figure 7-8. Estimated Cost of Other Site-Specific Activities at Current and Future NPL Sites: Base Case, FY 2000–FY 2009 (1999$)

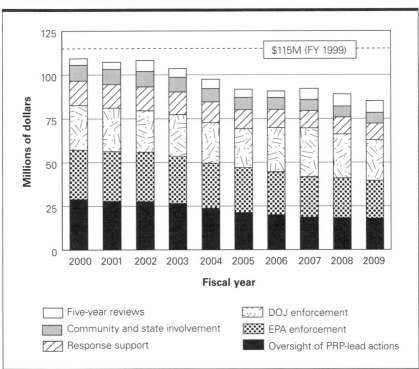

Total Cost of Other Site-Specific Activities

Overall, we estimate that for the base case, what we refer to as other site-specific costs for NPL sites—oversight, enforcement, response support, community and state involvement, and five-year reviews—will total approximately $975 million (or $1.09 billion, adjusted for inflation) from FY 2000 through FY 2009. Figure 7-8 shows these costs on an annual basis broken down by each activity in the base case. In the low case, we estimate that these costs will be about $924.0 million (or $1.03 billion, adjusted for inflation), and in the high case, about $1.04 billion (or $1.16 billion, adjusted for inflation).*

---

*Table H-6 in Appendix H shows the annual estimates for other site-specific costs from FY 2000 through FY 2009 under the three scenarios.

Cost of Non-NPL Sites

Although the focus of the remedial program is NPL sites, EPA also conducts response activities at sites proposed to the NPL, NPL equivalent sites, and other non-NPL sites where EPA determines some kind of response is warranted. These activities include site investigation, cleanup, and enforcement. Overall, EPA does not expend a significant part of its Superfund resources at these sites—a total of about $29.4 million (or 2% of total expenditures) in FY 1999. In that fiscal year, about $12.3 million of this total was spent on direct cleanup actions, mostly RI/FSs and RDs. The balance, $17.1 million, was spent on other site-specific activities—approximately $11.5 million for response support, $2.7 million for enforcement, $2.2 million for oversight, and $0.7 million for community and state involvement.[14]

In our estimates, we assume that EPA will continue to spend resources for activities at non-NPL sites. As we have no basis to assume a drastic change in these costs in either direction, we hold these costs constant at the FY 1999 level for each year from FY 2000 through FY 2009 under all three scenarios. Thus, over this period, we estimate that the cost to EPA of continuing to work at these non-NPL sites will be approximately $294 million (or $329 million, adjusted for inflation).

Total Remedial Program Cost

To summarize, we estimate that in the base case, EPA will incur approximately $6.55 billion (or $7.3 billion, adjusted for inflation) in total costs to implement the remedial program from FY 2000 through FY 2009, or $655 million on an average annual basis. This includes the cost of Fund-lead actions at current and future NPL sites; the cost of oversight of PRP-lead actions, enforcement, response support, community and state involvement, and five-year reviews; and the cost of response efforts at non-NPL sites. Figure 7-9 shows the relative contribution of each component of the remedial program.* In the low case, we estimate that total costs will be $5.71 billion (or $6.33 billion, adjusted for inflation), or about 13% less than the base case. In the high case, we estimate that the total will be $7.55 billion (or $8.43 billion, adjusted for inflation), or about 15% more. Thus, our annual estimates range from about $570.5 million to $754.8 million, compared with the $622.3 mil-

---

*See Table H-7 in Appendix H for the annual cost of each part of the remedial program from FY 2000 through FY 2009 under each scenario.

Figure 7-9. Estimated Cost of Remedial Program: Base Case, FY 2000–FY 2009 (1999$)

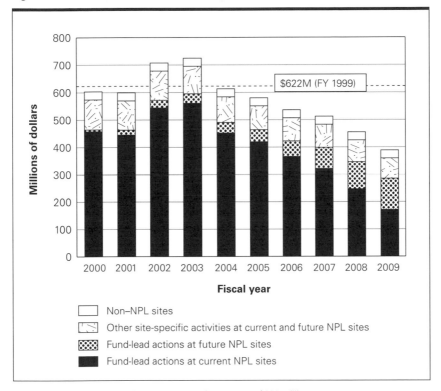

$622M (FY 1999)

Legend:
- Non–NPL sites
- Other site-specific activities at current and future NPL sites
- Fund-lead actions at future NPL sites
- Fund-lead actions at current NPL sites

Note: FY 1999 expenditures for these types of costs were $622 million.

lion spent by the Agency in FY 1999. This figure does not include other costs of running the Superfund program, such as the costs of rent, program management, and policy development. These costs are discussed—and added to the cost estimates—later in the chapter.

Cost of Site Assessment Activities

As noted in Chapter 5, before EPA determines whether a site needs a removal action or should be considered for—and ultimately listed on—the NPL, the Agency conducts a variety of site assessment activities. In total, site assessment expenses for a site that goes through all the steps of the assessment process range from about $80,000 to $180,000, depending on whether an expanded site inspection is necessary. Many sites go through only part of this process, since it is often found that the risks do not warrant further investigation. For this reason, we estimate the cost of site assessment activities on a programmatic rather than a site-specific level.

There is no reason to assume that EPA's investment in site assessment will change dramatically in the next decade, at least absent a major change in policy ordered by Congress or senior Agency management. The size and scope of site assessment work is largely a budget decision. For these reasons, we assume in our estimates that expenditures for site assessment will remain constant through FY 2009, at about $57.8 million annually, which was the amount spent in FY 1999. Of this total, 69% ($39.8 million) was spent on site-specific activities, 26% ($14.9 million) on site assessment program management and policy activities at EPA headquarters, and 5% ($3.1 million) for similar activities in the 10 EPA regional offices. Thus, from FY 2000 through FY 2009, we estimate that the cost to EPA of carrying out site assessment activities will be about $578 million (or $647 million, adjusted for inflation).

Cost of Program Staff, Management, and Support

Approximately $325.6 million, or 21%, of EPA's expenditures in FY 1999 were for what we characterize as program staff, management, and support. As noted in Chapter 6, these expenditures are often mischaracterized as nonsite costs, which suggests that they are not related to the cleanup of sites. In fact, many of the activities included in this category are essential to the Superfund cleanup programs, such as payroll and contract support. Some costs, such as those for the OSWER front office, although in theory discretionary, are to some degree necessary under any Superfund program. Some of these costs are fixed; others vary with the number of response actions at sites. For this reason, in our estimates of future costs, we hold a portion of program staff, management, and support costs constant during the FY 2000 through FY 2009 period and adjust others according to scaling factors.

We hold constant the annual cost of the following program staff, management, and support activities at their FY 1999 levels:

- $4 million for the OSWER front office;
- $2.3 million for the OECA front office and communications;
- $13.7 million for the National Enforcement Investigations Center;
- $31.4 million for state programs and grants;

- $4 million for the Office of Research and Development; and
- $3.1 million for the Office of General Counsel.

To estimate the future cost of remedial program policy, management, and staff, response action management, and laboratory support, we use the adjusted all-actions scaling factor. As described previously, this scaling factor totals the number of cleanup actions under way (adjusting for mega sites, LTRA, and PRP LR) at NPL sites at the end of each fiscal year to determine that year's workload.* We adjust these components of program staff, management, and support costs up or down from their FY 1999 levels, which were $113.9 million for remedial program policy, management and staff, $37.5 million for response action management, and $16.6 million for laboratory support.

To estimate the future cost associated with enforcement program policy, management and staff, we separate the activities into those related to NPL sites and those related to non-NPL sites. In FY 1999 the ratio of site-specific enforcement costs spent at NPL sites to non-NPL sites was 76% to 24%. We estimate the portion of the costs linked to NPL sites using the NPL activity scaling factor to adjust future costs in relation to FY 1999 levels. We hold constant the portion of the costs associated with non-NPL sites, which is consistent with how we estimate the cost of site-specific activities at non-NPL sites. Figure 7-10 displays the estimates for the base case of the components of program staff, management, and support costs.

In sum, for the base case, adding the portion of program staff, management, and support held constant to the portion adjusted by scaling factors yields a total cost estimate of about $2.69 billion (or $2.99 billion, adjusted for inflation) from FY 2000 through FY 2009. For the low case, the total estimate is $2.51 billion (or $2.78 billion, adjusted for inflation), and for the high case, $2.89 billion (or $3.23 billion, adjusted for inflation). Thus, the range of average annual costs is $250.9 million to $289.2 million, compared with the $325.6 million EPA spent in FY 1999.

Cost of Program Administration

As discussed in Chapter 6, numerous general administration costs are necessary to run the Superfund program. These costs cover the following:

---

*A small portion of the cost of remedial program policy, management, and staff activities is likely associated with non-NPL sites.

Figure 7-10. Estimated Cost of Program Staff, Management, and Support: Base Case, FY 2000–FY 2009 (1999$)

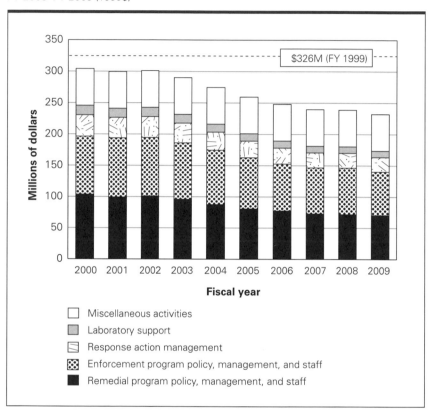

Notes: "Miscellaneous activities" includes all other program staff, management, and support activities: OSWER front office, OECA front office and communications, National Enforcement Investigations Center, state programs and grants, the Office of Research and Development, and the Office of General Counsel. We hold the annual cost of these activities constant at their level of FY 1999 expenditures.

FY 1999 expenditures for these types of costs were $326 million.

- rent, facilities operation and maintenance, and equipment and supplies;
- activities performed by staff in the Office of Administration and Resources Management (OARM), including contract and grant management, human resources management, and information systems and services, as well as regional administrative and resource management staff time; and
- activities performed by staff in the Office of the Chief Financial Officer (OCFO), such as financial management and budget and strategic planning.

In FY 1999 program administration costs totaled $114.1 million, or about 7% of overall Superfund expenditures—$61.2 million for rent, facilities operation and maintenance, and equipment and supplies, $34.7 million for activities carried out by OARM and for regional administrative and resource management staff, and $18.2 million for activities carried out by OCFO.*

We assume that the level of cost for these components of the Superfund program will ebb and flow with the overall size of the rest of the program, which we measure using a "total-dollars" scaling factor. To calculate this scaling factor, we sum our estimates for all the other costs of the Superfund program for each fiscal year, excluding those for the National Institute for Environmental Health Sciences (NIEHS), the U.S. Coast Guard, and the EPA Office of the Inspector General (OIG), and then adjust program administration costs (except for the cost of OARM information systems and services, which are held constant) accordingly, using FY 1999 as the base. We exclude NIEHS, the Coast Guard, and OIG because the Superfund program does not pick up their program administration costs, such as rent or facilities operation and maintenance.

For example, we estimate that the portion of Superfund costs associated with activities performed within OARM, along with costs for regional administrative and resource management staff time, will increase slightly during the first half of the FY 2000 through FY 2009 period. This increase reflects the rise in anticipated expenditures that is driven by the expected increase in the cost of Fund-lead actions at current NPL sites. From the total of $34.7 million in OARM expenditures in FY 1999, the model shows these costs peaking for the base case at about $37.9 million in FY 2003; these costs then gradually decrease, declining to about $29.5 million in FY 2009. The trends in expenditures are similar for the costs associated with activities that will be performed by OCFO staff, as well as for rent, facilities operation and maintenance, and equipment and supplies.

For the base case, we estimate that total program administration costs from FY 2000 through FY 2009 will be approximately $1.10 billion (or $1.23 billion, adjusted for inflation)—$589.8 million for rent, facilities operation and maintenance, and equipment and supplies, $335.3 million for OARM and regional

---

*Some of these costs should be apportioned to the brownfields and federal facilities programs within OSWER, but we had insufficient information to do so. As a result, our figures for program administration costs for the programs paid for with Trust Fund dollars are slightly higher than what is truly needed for the programs included as part of this study.

administrative and resource management staff time, and $175.4 million for OCFO. For the low case, we project that total program administration costs will be about $1.02 billion (or $1.13 billion, adjusted for inflation). For the high case, we estimate that total program administration costs will be about $1.20 billion (or $1.34 billion, adjusted for inflation).

Cost of Other Programs and Agencies

The final components of our estimates are for other programs and federal agencies that receive money from EPA's Superfund appropriation to finance their Superfund-related activities. As discussed in Chapter 6, included are programs within EPA—the Technology Innovation Office (TIO), the Chemical Emergency Preparedness and Prevention Office (CEPPO), and OIG—and other agencies, such as NIEHS and the Coast Guard. The level of some of these expenditures (such as TIO and CEPPO) is at the discretion of EPA, but Congress specifies the amounts that go to OIG, NIEHS, and the Coast Guard in its annual appropriations to the Agency.

We did not independently analyze these programs and agencies to determine whether their level of funding for Superfund-related efforts was appropriate or what level of expenditures would be needed in future years. Rather, in the model, we assume that these costs to EPA will remain constant at FY 1999 levels from FY 2000 through FY 2009. The FY 1999 expenditures for TIO, CEPPO, OIG, NIEHS, and the Coast Guard were $8.1 million, $7.6 million, $20 million, $60 million, and $4.8 million, respectively. From FY 2000 through FY 2009, the costs of these other programs and agencies to EPA would be just over $1 billion (or $1.12 billion, adjusted for inflation).

Total Estimated Cost of Superfund Program to EPA: Three Scenarios

Putting all the pieces together, we finally arrive at our estimates for the total future cost of the Superfund program to EPA. In the base case, we estimate that EPA's total cost to implement the Superfund program from FY 2000 through FY 2009 will be approximately $15.1 billion (or $16.9 billion, adjusted for inflation). Our estimates suggest that there will be a small decrease in resource needs in FY 2000 and FY 2001, compared with FY 1999 expenditures of just under $1.54 billion, and then an increase over the next two fiscal years, FY 2002 and FY 2003. In FY 2004 and FY 2005, our model indicates that EPA will need about the same amount of funds as in FY 1999, and from FY 2006 forward, the Agency will require a gradually declining level of resources to implement the Superfund program. Table 7-4 shows the cost for each of the six program elements for the base case for each year from FY 2000 through FY 2009.

Our low- and high- case estimates provide the likely range of future costs of the Superfund program. In the low case, we estimate that the total cost from FY 2000 through FY 2009 will be approximately $14 billion (or $15.6 billion, adjusted for inflation), about 7.3% less than in the base case. In the high case, the total cost is estimated at about $16.4 billion (or $18.3 billion, adjusted for inflation), approximately 8.6% more than in the base case. Figure 7-11 shows the total annual cost under each scenario.*

Although the total 10-year cost of the low case is only about 15% less than the high case, the difference is more pronounced on an annual basis. By FY 2009 the estimated cost in the low case is approximately $354 million (or 24%) less than in the high case. Between FY 2003 and FY 2009, the annual difference between the estimated future cost for the low case and the high case averages about $300 million.

It is worth noting that annual congressional appropriations for Superfund have historically not increased to take into account inflation. Thus, each year as federal salaries rise with cost-of-living adjustments and the cost of goods and services purchased by EPA increases, the actual Superfund budget for cleanup gets smaller. Table 7-5 presents the total estimates (in 1999 and inflation-adjusted dollars) under the three scenarios, breaking each scenario down into the six elements of the Superfund program.

Conclusions

One of the primary motivations for this report was to determine when the Superfund program was going to "ramp down." The estimates presented in this chapter should help answer this question. Our estimates do *not* show a ramp-down of the Superfund program in the short term. In fact, even if no additional sites are listed on the NPL after FY 1999, under the base-case

---

*See Table H-8 in Appendix H for the total estimated annual costs of the six main program elements from FY 2000 through FY 2009 under each scenario and Table H-9 in Appendix H for these same estimates in inflation-adjusted dollars.

Table 7-4. Estimated Total Annual Cost to EPA of Superfund Program: Base Case, FY 2000–FY 2009 (Millions of 1999$)

Program	2000	2001	2002	2003	2004	2005	2006	2007	2008	2009
Removal program	317.8	317.8	317.8	317.8	317.8	317.8	317.8	317.8	317.8	317.8
Remedial program	605.8	617.0	756.3	813.5	711.3	679.4	638.1	627.6	579.3	521.7
Site assessment	57.8	57.8	57.8	57.8	57.8	57.8	57.8	57.8	57.8	57.8
Program staff, management, and support	304.2	299.5	301.2	290.1	274.7	259.5	248.0	239.7	239.0	232.1
Program administration	108.8	109.4	121.3	125.1	115.4	111.5	106.9	105.1	101.2	96.0
Other programs and agencies	100.5	100.5	100.5	100.5	100.5	100.5	100.5	100.5	100.5	100.5
Total	**1,495.0**	**1,502.1**	**1,654.8**	**1,704.8**	**1,577.5**	**1,526.6**	**1,469.1**	**1,448.5**	**1,395.6**	**1,326.0**

Note: Totals may not add exactly because of rounding.

Figure 7-11. Estimated Total Annual Cost to EPA of Superfund Program: Three Scenarios, FY 2000–FY 2009 (1999$)

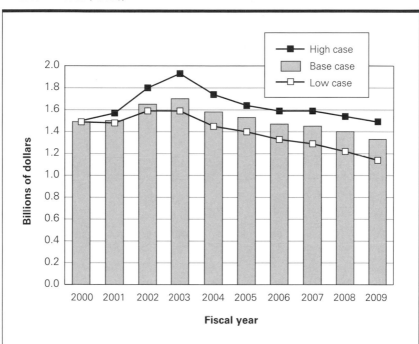

scenario EPA will need at least $1.4 billion—and sometimes more—each year to implement the Superfund program through FY 2004.

Figure 7-12 illustrates this point by separating the dollars that can be attributed to the current NPL under the base-case scenario (including the cost of the removal program and the other costs of running the Superfund program) from those that can be attributed solely to addressing future NPL sites. The shaded area in the figure represents the cost, in the base case, for the Superfund program assuming that *no* sites are added to the NPL from FY 2000 through FY 2009. If no new sites were added to the NPL after FY 1999, a ramp-down of program costs would begin around FY 2004, when annual expenditures begin to fall steadily.

Of course, there is no indication that EPA and the states will cease listing sites on the NPL. In fact, the number of NPL listings started to increase in FY 1999, when more sites (37) were listed than in any year since 1990. In FY 2000,

Table 7-5. Estimated Total Cost to EPA of Superfund Program: Three Scenarios, FY 2000–FY 2009 (Millions of Dollars)

	Base case		Low case		High case	
	1999$	Inflation-adjusted	1999$	Inflation-adjusted	1999$	Inflation-adjusted
Removal program	3,180	3,560	3,180	3,560	3,180	3,560
Remedial program	6,550	7,300	5,710	6,330	7,550	8,430
Site assessment	580	650	580	650	580	650
Program staff, management, and support	2,690	2,990	2,510	2,780	2,890	3,230
Program administration	1,100	1,230	1,020	1,130	1,200	1,340
Other programs and agencies	1,010	1,120	1,010	1,120	1,010	1,120
Total	**15,100**	**16,850**	**13,990**	**15,580**	**16,400**	**18,330**

Note: Totals may not add exactly because of rounding.

the upward trend continued with the addition of 36 sites (including 2 mega sites) to the NPL. We have estimated in the three scenarios discussed in this chapter that between 23 and 49 final sites will be added to the NPL each year from FY 2001 through FY 2009.

Figure 7-12 shows the estimated costs that can be attributed solely to sites added to the NPL after FY 1999, on top of the rest of the Superfund program under the base-case scenario. The figure also shows that the difference among the three future NPL scenarios is less than $140 million through FY 2006. By FY 2009, however, the difference in estimated costs is greater—$320 million.

Even though the sites added to the NPL from FY 2000 through FY 2009 will not account for the lion's share of the costs EPA will likely incur for the Superfund program during this decade, they nevertheless have considerable cost implications. Our model clearly shows that the listing decisions made today will have a real impact 10 years from now, since even cleanups at a site with only one operable unit will on average take approximately 11 years to be completed.

As has been mentioned repeatedly in this report, predicting exactly which sites are likely to be added to the NPL, how many among them will be mega

Figure 7-12. Estimated Total Annual Cost to EPA of Superfund Program: Current NPL Base Case with Three Future NPL Scenarios, FY 2000–FY 2009 (1999$)

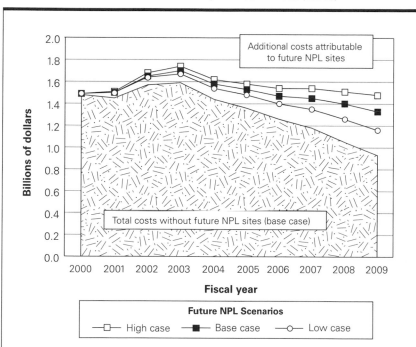

sites, and how many sites will have Fund-lead remedial actions is fraught with uncertainty. Those wanting to forecast the need for future Superfund dollars on a national level will need to scrutinize the sites added to the NPL each year and anticipate which sites are likely to be cleaned up with Trust Fund dollars and how much they will cost.

Appendix A

EPA Regional Offices

Region 1 Connecticut, Maine, Massachusetts, New Hampshire, Rhode Island, Vermont
Region 2 New Jersey, New York, Puerto Rico, U.S. Virgin Islands
Region 3 Delaware, Maryland, Pennsylvania, Virginia, West Virginia, District of Columbia
Region 4 Alabama, Florida, Georgia, Kentucky, Mississippi, North Carolina, South Carolina, Tennessee
Region 5 Illinois, Indiana, Michigan, Minnesota, Ohio, Wisconsin
Region 6 Arkansas, Louisiana, New Mexico, Oklahoma, Texas
Region 7 Iowa, Kansas, Missouri, Nebraska
Region 8 Colorado, Montana, North Dakota, South Dakota, Utah, Wyoming
Region 9 Arizona, California, Hawaii, Nevada, Guam, American Samoa
Region 10 Alaska, Idaho, Oregon, Washington

FY 2000 Status of 52 Sites Proposed to the NPL as of the End of FY 1999

Region and State		Site name	NPL proposal date	Status as of end of FY 2000
1	MA	General Electric - Housatonic River	9/25/97	NPL equivalent
2	NJ	Iceland Coin Laundry Area Ground Water Contamination	7/22/99	Final on NPL, 10/22/99
2	NJ	Lightman Drum Company	7/22/99	Final on NPL, 10/22/99
2	NJ	Route 561 Dump	7/28/98	Listing decision pending
2	NY	Peter Cooper Corporation (Markhams)	4/23/99	Final on NPL, 2/4/00
3	MD	68th Street Dump/Industrial Enterprises	1/19/99	Listing decision pending
3	PA	East Tenth Street	1/18/94	Listing decision pending, state-lead site
3	PA	Old Wilmington Road Groundwater Contamination	7/22/99	Final on NPL, 2/4/00
3	PA	Salford Quarry	4/1/97	Listing decision pending
3	WV	Vienna Tetrachloroethene	4/23/99	Final on NPL, 10/22/99
4	FL	Normandy Park Apartments	2/13/95	NPL equivalent
4	GA	Terry Creek Dredge Spoil Areas/ Hercules Outfall	4/1/97	NPL equivalent
4	MS	Chemfax, Inc.	6/23/93	Listing decision pending
4	MS	Potter Co.	5/10/93	Listing decision pending

Region and State		Site name	NPL proposal date	Status as of end of FY 2000
4	NC	Georgia-Pacific Corporation Hardwood Sawmill	1/19/99	Final on NPL, 10/22/99; removed from NPL by court action, 9/28/00
4	TN	American Bemberg	9/25/97	Listing decision pending[a]
5	IL	Circle Smelting Corp.	6/17/96	NPL equivalent candidate
5	IL	Evergreen Manor Ground Water Contamination	7/28/98	NPL equivalent
5	IL	Indian Refinery-Texaco Lawrenceville	7/28/98	Listing decision pending[b]
5	IL	Sauget Area 1	6/17/96	Listing decision pending
5	IN	U.S. Smelter and Lead Refinery, Inc.	2/7/92	Formal deferral to RCRA
5	MI	Bay City Middlegrounds	2/13/95	Listing decision pending, state-lead site
5	OH	Diamond Shamrock Corp. (Painesville Works)	5/10/93	Listing decision pending, state-lead site
5	OH	Dover Chemical Corp.	5/10/93	NPL equivalent
5	WI	Fox River NRDA/PCB Releases	7/28/98	NPL equivalent candidate
6	LA	Gulf State Utilities-North Ryan Street	2/13/95	NPL equivalent
6	LA	Highway 71/72 Refinery (Old CITGO Refinery-Bossier City)	2/13/95	NPL equivalent
6	NM	Fruit Avenue Plume	7/22/99	Final on NPL, 10/22/99
6	OK	National Zinc Corp.	5/10/93	Formal state deferral
6	TX	Garland Creosoting	7/22/99	Final on NPL, 10/22/99
6	TX	Star Lake Canal	7/22/99	Final on NPL, 7/27/00
6	TX	State Road 114 Ground Water Plume	7/22/99	Final on NPL, 10/22/99
7	IA	Waterloo Coal Gasification Plant	10/14/92	NPL equivalent
7	MO	Newton County Wells	1/19/99	Final on NPL, 7/27/00
8	CO	ASARCO, Inc. (Globe Plant)	5/10/93	NPL equivalent
8	CO	Smeltertown	2/7/92	NPL equivalent
8	MT	Basin Mining Area	7/22/99	Final on NPL, 10/22/99

Region and State		Site name	NPL proposal date	Status as of end of FY 2000
8	MT	Burlington Northern Livingston Shop Complex	8/23/94	NPL equivalent
8	MT	Upper Tenmile Creek Mining Area	7/22/99	Final on NPL, 10/22/99
8	UT	International Smelting and Refining	4/23/99	Final on NPL, 7/27/00
8	UT	Jacobs Smelter	7/22/99	Final on NPL, 2/4/00
8	UT	Kennecott (North Zone)	1/18/94	NPL equivalent
8	UT	Kennecott (South Zone)	1/18/94	NPL equivalent
8	UT	Murray Smelter	1/18/94	NPL equivalent
8	UT	Richardson Flat Tailings	2/7/92	NPL equivalent
9	CA	Cooper Drum Co.	2/7/92	Listing decision pending
9	CA	GBF, Inc., Dump	2/7/92	Listing decision pending, state-lead site
9	CA	Stoker Company	7/29/91	Listing decision pending
10	ID	Blackbird Mine	5/10/93	NPL equivalent
10	ID	Triumph Mine Tailings Piles	5/10/93	Formal state deferral
10	OR	East Multnomah County Ground Water Contamination	5/10/93	Listing decision pending, state-lead site
10	WA	Midnite Mine	2/16/99	Final on NPL, 5/1/00

[a]Removed from proposed NPL, 11/29/2000.

[b]Final on the NPL, 12/1/00.

Source: Data provided to RFF by EPA, October 2000.

Appendix C

Data Sources

Main Datasets

In March 2000, the U.S. Environmental Protection Agency (EPA) provided RFF with four large datasets derived from two of the Agency's databases: the Comprehensive Environmental Response, Compensation, and Liability Information System (CERCLIS), which is EPA's major Superfund action tracking system, and the Integrated Financial Management System (IFMS), which is EPA's internal financial accounting system. All the datasets contain data from the beginning of the program through the end of FY 1999. We received two datasets with CERCLIS data—a site-level dataset and an action-level dataset. As discussed below, the site-level dataset contains descriptive information about National Priorities List (NPL) sites, and the action-level dataset contains descriptive information about past and present actions at these NPL sites. The CERCLIS datasets do not include information on NPL sites that are federal facilities—that is, where the site is being cleaned up by a federal agency other than EPA, such as the Department of Energy or the Department of Defense.* RFF also received two sets of IFMS data, one with historical obligation data and the other with historical expenditure data.

RFF Site-Level Dataset

The CERCLIS site-level dataset provided by EPA contains information for each of the 1,297 nonfederal NPL sites that were proposed, final, and/or deleted at the end of FY 1999. The dataset includes information for each site,

---

*Sites where a federal agency is a (but not the sole) responsible party are included.

such as the CERCLIS identification number (CERCLIS ID), site/spill identification number (SSID), site name, region, state, and congressional district. Each site entry has variables for its NPL status and the date the site was proposed to, listed as final on, and/or deleted from the NPL. Other variables include construction-complete status and, when applicable, the date of construction completion, the number of operable units, and site area.

From the CERCLIS site-level dataset and other data sources, we created the RFF site-level dataset by adding new variables and changing others (for example, we recategorize action lead information, as discussed in Appendix F). Using information collected from the 10 EPA regions, we label each site as either a mega or a nonmega site. Mega sites are defined as those sites with total or expected removal and remedial costs of $50 million or more. We also use a variable for teenager and nonteenager sites. A teenager site is a final site that was proposed to the NPL before FY 1987 and was not construction-complete as of the end of FY 1999. We also classify the 1,297 sites according to 11 RFF site types, described in detail in Appendix F. The final RFF site-level dataset has 1,297 entries with 42 variables. For most of our analyses, we use only information on the 1,245 final and deleted NPL sites; the remaining 52 sites were proposed but not final on the NPL at the end of FY 1999, the cutoff date for most of our data.

RFF Action-Level Dataset

The CERCLIS action-level dataset provided by EPA contains information for each action completed or under way at proposed, final, and deleted NPL sites as of the end of FY 1999. Actions refer to responses taken at the site, such as a removal action, a remedial investigation/feasibility study (RI/FS), a remedial design (RD), a remedial action (RA), or a long-term response action (LTRA), and are reported at the operable unit level. The database contains descriptive information for each action, such as the CERCLIS ID and the name of the site, as well as the operable unit number at which it was conducted. Each entry also has a variable indicating the lead—that is, whether the action was funded by EPA (Fund-lead) or by the potentially responsible party (PRP-lead)—using one of EPA's classifications to designate leads. Finally, the CERCLIS action-level dataset also includes the date of the action's start and completion.

EPA's original action-level dataset did not include information about future actions to be conducted at a site, referred to as EPA planning data, such as the planned completion date for actions that were still under way at the end of FY 1999. At RFF's request, in May 2000, EPA provided a supplemental dataset derived from CERCLIS containing Agency planning data. This dataset includes the planned completion dates for all actions that were under way at the end of FY 1999 as well as the start date of the next planned action for each operable unit. Each next planned action is also characterized by its expected lead. EPA considers planning data "enforcement confidential," meaning that RFF cannot make these data public. The Agency cautioned us that these data are not necessarily accurate, as they are based on remedial project managers' estimates of when future actions will be conducted—often years in advance of actual implementation. As a result, the estimates tend to be optimistic and are not necessarily informed by budget realities. For this reason, we decided to ask EPA only for near-term data, which are likely to be more reliable.

With the CERCLIS action-level dataset and the supplemental data provided by the Agency, we created the RFF action-level dataset. For each action in the RFF action-level dataset, we recode the lead information provided by EPA to categories more amenable to our analyses. These modifications, explained in Appendix F, simplify the coding to separate those actions paid for by the Trust Fund from those that are not. For example, for our purposes, an action paid for by another federal agency (which can happen even at a site not formally deemed a federal facility) is the same as an action paid for by a responsible party, since the funds for cleanup do not come out of the Trust Fund.

EPA's initial action-level dataset also did not include information regarding no further action (NFA) records of decision (RODs). EPA provided RFF supplemental data in April and May 2000 that indicated which RODs resulted in NFA determinations, which we have added to our database.

Lastly, we incorporate many of the variables from the RFF site-level dataset into the RFF action-level dataset (e.g., mega site, teenager site). The final dataset has 12,739 entries with 70 variables for each of the 1,297 nonfederal proposed, final, and deleted NPL sites.

RFF IFMS Dataset

EPA provided RFF with both obligation and expenditure data from IFMS, but we use only the expenditure data in our analyses. The IFMS expenditure dataset contains one entry for each disbursement made by EPA from the beginning of the Superfund program through the end of FY 1999. Our IFMS data include two major types of expenditures—those charged directly to specific sites and those charged to other program activities. We use the site-specific expenditure data to calculate the average cost of remedial pipeline actions and to estimate the cost of site-specific support activities. We use nonsite-specific data from FY 1999 to provide a baseline of all remaining program costs. Each type of expenditure data is described below.

Each site-specific entry in the RFF IFMS dataset has an associated SSID, which is used to identify the specific site (NPL or non-NPL) the expenditure is associated with. For site-specific expenditures made at NPL sites, we mapped these SSIDs to the corresponding CERCLIS IDs in the RFF site-level dataset and the RFF action-level dataset. Each site-specific expenditure in the RFF IFMS dataset is also categorized by two major variables—the type of response action and the category of expenditure. The designations for type of response action include those for direct cleanup actions (removal actions, RI/FSs, RDs, and RAs), site assessment, post-construction activities (LTRAs and five-year reviews), and other site-specific activities (oversight, enforcement, response support, and state and community involvement). There are also two other types of site-specific expenditures in the dataset—operation and maintenance, and groundwater monitoring—which we merged with response support because they were small.

Each site-specific expenditure is also placed into one of two categories: extramural or intramural. Extramural expenditures refer to payments made for activities conducted outside EPA through contracts, interagency agreements, or cooperative agreements with states. Intramural expenditures refer to payments for activities within the Agency, such as payroll, Department of Justice (DOJ),* or travel and other miscellaneous expenses (e.g., supplies). Each expenditure is also characterized by the fiscal year in which it was made.

Expenditures that are not attributable to sites, so-called nonsite expenditures, are organized in the RFF IFMS dataset in the same response action categories as site-specific expenditures. We worked with EPA to convert these nonsite expenditures into categories more useful for understanding the costs associated with other Superfund program activities. These expenditures have been reassigned into three categories—response support, enforcement, and management and support—and then characterized by whether they were made at the regional or headquarters level. To further characterize the expenditures within each category, we consulted staff at EPA in the Office of Emergency and Remedial Response for response support, the Office of Enforce-

---

*For the purposes of analysis, we include DOJ expenditures under intramural costs, even though the money is being spent outside EPA.

ment and Compliance Assurance for enforcement, and the Office of Administration and Resources Management and the Office of the Chief Financial Officer for management and support.

The RFF IFMS dataset also contains expenditure data for the brownfields program and for activities at federal facilities, but we do not use these data because these components of the Superfund program are excluded from our study.

Other Datasets

In addition to the CERCLIS and IFMS datasets described above, we received from EPA a number of smaller datasets, described below, with historical, cost, and descriptive information on removal actions and five-year reviews. EPA also provided RFF with other data on specific topics throughout the duration of the study.

Removal Action Datasets

In March 2000, EPA gave RFF a dataset with historical smmary data regarding the numbers and types—emergency response, time-critical actions, non-time-critical actions, and others—of removal actions undertaken from FY 1980 through FY 1999 in each of the 10 EPA regions. This dataset, referred to as removals historical summary data, is derived from CERCLIS and contains information about 5,297 removal actions. For the FY 1992 through FY 1999 period, which we use to characterize recent historical removal actions by category and EPA region, there were 2,543 removal actions for which we have complete data (this does not include 93 additional removal actions coded "other").

EPA also provided RFF with another CERCLIS dataset in March 2000 relating to removal actions conducted from FY 1980 through FY 1999. From these data, RFF has created a dataset of all removal actions with "valid removal starts" conducted at nonfederal sites from FY 1992 through FY 1999. We define a *valid removal start* as an action for which a start date is recorded in CERCLIS and no other action has been taken that would call that start date into question (e.g., takeover by a responsible party, transfer to another agency, or other action constituting an "anomaly" in CERCLIS). We exclude 237 removal actions because of missing SSIDs, which link CERCLIS information to the financial information in IFMS. The dataset we use in many of our analyses of the removal program is a subset of these data and includes 2,537 removal action starts representing 2,053 sites initiated from FY 1992 through FY 1999. We refer to this dataset as removal action starts data. In some analyses, we had to pare down the dataset further because data were missing from relevant fields (e.g., removal type data for analysis of lead according to type of removal).

We received a third dataset from EPA in April 2000 pertaining to removal actions. Derived from the Emergency Response Notification System, an EPA database that compiles incident notifications from various sources, these data have information on historical release notifications (such as reports of oil spills and hazardous materials releases) received by the Agency.

Five-Year Review Datasets

In March 2000, EPA provided RFF with three datasets derived from the Agency's five-year review program implementation and management system, with information regarding five-year reviews as of February 18, 2000. The datasets include information on federal facilities, but we do not use information on these sites in our analyses. The first dataset contains historical information on the five-year reviews completed through the first quarter of FY 2000. The second dataset lists the five-year reviews that EPA planned to complete from FY 2000 through FY 2002. The last dataset indicates whether a five-year review is required by statute or policy or not required at all for 1,095 sites, 982 of which are listed as final or deleted on the NPL.

Appendix D

RFF Data Collection

To supplement the data provided by the U.S. Environmental Protection Agency (EPA) (described in detail in Appendix C), we collected new data from the Agency. We focused our data collection efforts in three areas: mega sites (those sites with total or expected removal and remedial action costs of $50 million or more), Fund-lead remedial actions at National Priority List (NPL) sites, and the removal program. Each is discussed below. We also conducted interviews with the 10 EPA regional Superfund division directors and officials from 9 states to inform our estimates of the number and character of sites likely to be added to the NPL each year from FY 2000 through FY 2009; the interviews are described in Appendix E.

Mega Sites

Our analyses of EPA expenditures, coupled with the findings of a previous study of Superfund costs conducted by the Congressional Budget Office, clearly show that certain sites on the NPL cost considerably more to clean up than others. For this reason, we decided to collect site-specific information for the most expensive sites on the NPL—mega sites—to improve our estimates of the future cost of the Superfund program. Mega sites may have both PRP- and Fund-lead actions. At our request, the 10 EPA regional offices identified 112 NPL sites—109 listed as final and 3 listed as deleted—that, as of the end of FY 1999, met this definition. These sites are listed in Table D-1. The Agency also identified an additional 7 mega sites that were listed as proposed to the NPL at the end of FY 1999. These sites are General Electric–Housatonic River in Massachusetts (Region 1), Sauget Area 1 in Illinois (Region 5),

Table D-1. Mega Sites on the Current NPL by Region, End of FY 1999

Region 1 (7)

- Baird & McGuire, MA
- Industri-Plex, MA
- New Bedford, MA
- Nyanza Chemical Waste Dump, MA
- Raymark Industries, Inc., CT
- Silresim Chemical Corp., MA
- Wells G&H, MA

Region 2 (20)

- American Cyanamid, Co., NJ
- Bridgeport Rental and Oil Services, NJ
- Ciba-Geigy Corp. (Toms River), NJ
- Diamond Alkali Co., NJ
- Federal Creosote, NJ
- GEMS Landfill, NJ
- General Motors (Central Foundry Division), NY
- Glen Ridge Radium, NJ
- Helen Kramer Landfill, NJ
- Hooker (Hyde Park), NY
- Hooker (S Area), NY
- Hudson River PCBs, NY
- Lipari Landfill, NJ
- Love Canal, NY
- Marathon Battery Corp., NY
- Montclair/West Orange Radium, NJ
- Onondaga Lake, NY
- U.S. Radium, Corp., NJ
- Vega Baja Solid Waste Disposal, Puerto Rico
- Vineland Chemical Company, Inc., NJ

Region 3 (12)

- Austin Avenue Radiation, PA
- Avtex Fibers, Inc., VA
- Drake Chemical, PA
- E.I. DuPont De Nemours & Co. (Newport Landfill), DE
- Fike Chemical, Inc., WV
- Heleva Landfill, PA
- MW Manufacturing, PA
- Palmerton Zinc Pile, PA
- Saltville Waste Disposal Ponds, VA
- Southern Maryland Wood Treating, MD
- Tysons Dump, PA
- Whitmoyer Laboratories, PA

Region 4 (1)

- Interstate Lead Company (ILCO), AL

Region 5 (14)

- Allied Chemical & Ironton Coke, OH
- Allied Paper/Portage Creek/Kalamazoo River, MI
- American Chemical Services, Inc., IN
- Big D Campground, OH
- Continental Steel Corp., IN
- Kerr-McGee (Cress Creek/West Branch of DuPage River), IL
- Kerr-McGee (Reed-Keppler Park), IL
- Kerr-McGee (Residential Areas), IL
- NL Industries/Taracorp Lead Smelter, IL
- Ott/Story/Cordova Chemical Co., MI
- Ottawa Radiation Areas, IL
- Sheboygan Harbor & River, WI
- Southeast Rockford Ground Water Contamination, IL
- St. Louis River, MN

Region 6 (14)

- Alcoa (Point Comfort)/Lavaca Bay, TX
- AT&SF (Albuquerque), NM
- Bayou Bonfouca, LA
- Brio Refining, Inc., TX
- Cleve Reber, LA
- French, Ltd., TX
- Geneva Industries/Fuhrmann Energy, TX
- Hardage/Criner, OK
- MOTCO, Inc., TX
- Petro-Processors of Louisiana, Inc., LA
- Sand Springs Petrochemical Complex, OK
- Sikes Disposal Pits, TX
- Tar Creek (Ottawa County), OK
- Vertac, Inc., AR

Region 7 (5)

- Big River Mine Tailings/St. Joe Minerals Corp., MO
- Cherokee County, KS
- Hastings Ground Water Contamination, NE
- Oronogo-Duenweg Mining Belt, MO
- Times Beach, MO

continued on next page

Table D-1. *continued*

Region 8 (10)

- Anaconda Co. Smelter, MT
- California Gulch, CO
- Denver Radium, CO
- Eagle Mine, CO
- Lowry Landfill, CO
- Milltown Reservoir Sediments, MT
- Silver Bow Creek/Butte Area, MT
- Summitville Mine, CO
- Uravan Uranium Project (Union Carbide Corp.), CO
- Whitewood Creek, SD

Region 9 (24)

- Aerojet General Corp., CA
- Fairchild Semiconductor Corp (Mountain View), CA
- Indian Bend Wash Area, AZ
- Intel Corp. (Mountain View Plant), CA
- Iron Mountain Mine, CA
- J.H. Baxter & Co., CA
- Litchfield Airport Area, AZ
- McColl, CA
- McCormick & Baxter Creosoting Co., CA
- Montrose Chemical Corp., CA

- Motorola, Inc. (52nd Street Plant), AZ
- Newmark Ground Water Contamination, CA
- Omega Chemical Corp., CA
- Operating Industries, Inc. Landfill, CA
- Raytheon Corp., CA
- San Fernando Valley (Area 1), CA
- San Fernando Valley (Area 2), CA
- San Gabriel Valley (Area 1), CA
- San Gabriel Valley (Area 2), CA
- San Gabriel Valley (Area 4), CA
- Southern California Edison Co. (Visalia Poleyard), CA
- Stringfellow, CA
- Sulphur Bank Mercury Mine, CA
- Tucson International Airport Area, AZ

Region 10 (5)

- Bunker Hill Mining & Metallurgical Complex, ID
- Commencement Bay-Nearshore Tideflats, WA
- Commencement Bay-South Tacoma Channel, WA
- Western Processing Co., Inc., WA
- Wyckoff Co./Eagle Harbor, WA

Indian Refinery–Texaco Lawrenceville in Illinois (Region 5), Fox River NRDA/PCB Releases in Wisconsin (Region 5), Kennecott (South Zone) and Kennecott (North Zone) in Utah (Region 8), and Midnite Mine in Washington (Region 10).

RFF developed a questionnaire (the RFF mega site survey) to collect site-specific information on the 112 mega sites. The questionnaire was reviewed in draft by EPA staff and outside Superfund experts. The Office of Solid Waste and Emergency Response distributed the questionnaire in August 2000 directly to the remedial project manager (RPM) responsible for each mega site. Along with the survey, the RPMs received detailed instructions, a summary of past EPA extramural expenditures [based on Integrated Financial Management System (IFMS) expenditure data through FY 1999] at the site, and the current status of all operable units at the site (based on data from EPA's Comprehensive Environmental Response, Compensation, and Liability Information System through FY 1999). To facilitate the administration of the survey, RFF held a conference call with regional office staff assigned by the Agency to serve as liaisons to answer RPMs' questions about the survey. EPA also coordinated three sessions during which RFF staff answered questions directly from the RPMs.

The RPMs were asked to identify all ongoing and future removal actions,* remedial investigation/feasibility studies (RI/FSs), remedial designs (RDs), remedial actions (RAs), and long-term response actions (LTRAs) at their sites. The RPMs were also asked whether each action was (or was likely to be) Fund-lead or PRP-lead. If Fund-lead, we asked the RPM to estimate the amount of extramural Trust Fund dollars that would be needed for each action, by year, from FY 2000 through FY 2009. For future PRP-lead actions, we asked the RPMs to estimate the starting and ending fiscal year of each action, as well as the amount, if any, of extramural Trust Fund dollars that would be spent as part of the action.

In addition to using the survey data to estimate the cost of cleanup actions at the 112 mega sites, we also use the information about the number of actions projected to be under way and the number of actions not yet started to forecast the number of total actions under way in each fiscal year from FY 2000 through FY 2009. From this information, we are able to take into account the workload generated by both Fund-lead and PRP-lead actions at mega sites. This information is used to "load" the costs of other site-specific activities (e.g., oversight, response support) and other parts of the Superfund program (e.g., management and support), as explained in Appendix G.

With the support of EPA, we received a 100% response rate to the RFF mega site survey. EPA reviewed the completed surveys for quality control purposes, and RFF also reviewed each response, resolving all questions and uncertainties through followup discussions and emails with individual RPMs and other EPA regional staff.

Fund-Lead Remedial Actions

Most past analyses of the cost of Superfund cleanups have used estimates from record of decision (ROD) documents. Because we were concerned about the reliability of these data, we conducted a survey in May 2000 to collect data on the cost of implementing the remedial actions that were to follow 158 Fund-lead remedial designs (RDs) completed from FY 1995 through FY 1999 at 121 nonfederal NPL sites. We chose this sample to reflect recent trends in remedy selection. The 158 RDs represented all 11 RFF site types and

---

*We do not use the RPMs' estimates of future removal costs in our model, since we decided to base our future estimate of the removal program on current overall removal program expenditures.

both mega and nonmega sites. We were seeking a check on the remedial action unit costs we calculated from IFMS expenditure data (discussed in Appendix F) and data on LTRA frequency and costs.

For each RD, we asked the RPM overseeing the action to report the cost of the RA designed in the RD in terms of the value of the accepted RA contract bid. This value reflects the actual amount EPA agreed to pay the contractor to implement the RA. We also asked the RPMs to provide additional information about each RA, including the media or area it addressed (e.g., soil, groundwater), the selected remedy, the date of the contract (used to adjust the estimates into constant 1999 dollars), whether the contract was for the entire RA at the operable unit, the amount of the contract that was for construction and its expected duration, and the amount of the contract that was for LTRA (if applicable) and its expected duration. We use the data collected in the survey to create the RFF remedial action cost database.

Although we received responses from RPMs in all 10 EPA regional offices, many of the responses were incomplete or failed to meet our requirements for other reasons and are not included in the RFF remedial action cost database.* We also combine response data when there are multiple remedial actions at a single operable unit, which is consistent with our focus on the operable unit level rather than action level in all of our analyses. The final dataset includes at least partial information for 112 remedial actions.

We use the data in the RFF remedial action cost database for two purposes. First, we calculate the average cost of remedial actions, based on the value of the remedial action contracts. We have complete information for 70 remedial actions at 59 sites. Table D-2 summarizes the costs by all sites, mega sites, nonmega sites, and site type (for the nonmega sites only) for which contract awards are known. The small sample size for many of the site types brings into question site type–specific averages, but on an aggregate level, the estimates of $29.53 million for a single remedial action at a mega site and $5.44 million for a remedial action at a nonmega site are comparable to the estimates we calculate using historical expenditure data in IFMS.

We also use the RFF remedial action cost database to examine the frequency of LTRAs—that is, the number of RAs followed by LTRAs—and their average annual cost. We have complete data for 107 of the 112 remedial actions in the database. Of these 107 remedial actions, an LTRA is expected to follow at 36% (38). As we discuss in detail in Appendix F, from these data we calculate that the

---

*The main reasons responses were eliminated were that the actions reported did not represent the entire remedial action, or the response was for remedial actions outside the requested time frame.

Table D-2. Estimated Remedial Action Cost by Site Type

Type of site	Number of sites	Average cost (1999$)
All sites	70	7,500,000
Mega sites	6	30,000,000
Nonmega sites	64	5,400,000
Chemical manufacturing	4	1,700,000
Oil refining	1	9,400,000
Mining	3	18,000,000
Wood preserving	8	13,000,000
Coal gasification and other industrial	18	3,300,000
Recycling	5	3,400,000
Captive waste handling and disposal	4	4,700,000
Noncaptive waste handling and disposal	4	6,400,000
Transportation	2	13,000,000
Contaminated areas	8	1,200,000
Miscellaneous	7	2,000,000

Source: RFF remedial action cost database.

average annual LTRA cost is approximately $760,000 for all sites, $2 million for mega sites, and $450,000 for nonmega sites.

Removal Program

RFF interviewed the 10 regional program managers of the removal program in July 2000. Our initial goal was to obtain quantitative estimates of anticipated future removal actions in each of the 10 EPA regions. Although it quickly became apparent that the basis for such estimates did not exist, the interviews nonetheless proved valuable. The regional program managers provided a "ground-level" view of the removal program and explained its role within the overall Superfund program, the triage system that typifies the planning process in most EPA regions, and the factors that influence the number of sites requiring removal actions that the regions anticipate encountering. A copy of the standardized questionnaire appears as Figure D-1.

Figure D-1. RFF Regional Removal Questionnaire *(continues on next two pages)*

RESOURCES FOR THE FUTURE
REGIONAL REMOVAL QUESTIONNAIRE

Region _____ Branch Chief _____
Number of years in program _____

Questions for Regional Removal Branch Chiefs

1. What do you see to be the primary role(s) of the removals program in your region?
   a. Does this goal/Do these goals differ for removals at NPL sites and non-NPL sites?
   b. Are they different for different types of removals (emergency, time-critical, non-time-critical) If so, explain.

2. Describe your region's policy toward use of removals in the Superfund program.
   a. Are they used to keep sites off the NPL? Yes/No
   b. Do you consider them an integral part of the remediation program or a separate function?
   c. Do removal actions prevent or limit the need for future remediation?

3. For descriptive purposes, please give a typical example of an emergency removal, a non-time-critical removal, a time-critical removal that you encounter in your region. What kind of sites are we talking about here (new facilities, old industrial facilities)?

4. What are the sources of sites referred to you for removal actions? Name them (for example):
   a. Local fire/police/emergency management groups
   b. State superfund programs
   c. NPL
   d. Other _____

For each source, in what proportions are the different removal actions typically referred to you?

	Local fire/ emergency	State superfund	NPL	Other (specify)
Removal type				
Emergency				
Time-critical				
Non-time-critical				
All removals				

Figure D-1. RFF Regional Removal Questionnaire *(continued)*

In other words, are certain types of removal actions more likely to come from one source or another (e.g., emergencies from local fire or other emergency management agencies)?

Do you anticipate any change in this pattern in the next 5–10 years?

If so, what kind of changes do you anticipate? and what factors might contribute to them?

5. We are interested in understanding the process you undergo to plan the number of removals you do each year. Do you have formal meetings with the local, state, or federal entities from whom removals are referred to you? Yes/No.

   If so, with whom and how often (annually/quarterly/monthly/other)?

   If not, how are you notified of sites potentially needing removal actions? What is the process?

   What is the typical lead time?

   Break down by type of removal and type of site (NPL/non-NPL) as necessary.

6. What is your estimate of the potential "pool" (inventory) of sites that may require removal actions of some kind in the next year, 5 years, 10 years?
   a. State or other programs (non-NPL) and
   b. The Superfund program (NPL)

7. Basic budget questions:
   a. What are all the elements that are in your budget (e.g., extramural, intramural, travel, START, interagency agreements, etc.)?
   b. What is the total removals budget for your region given this definition (i.e., beginning fiscal year)?
   c. What is the breakdown in terms of the elements that you first defined for your budget?
   d. What is the total workforce (full-time equivalent, or FTE) assigned to the removal program in your region? How many of these are funded by Superfund, OPA?

8. If you could double the budget and FTE to perform removals, would you double the number of removals you do? Yes/No. If no, what other constraints are operating (e.g., the rate at which sites appear needing action, etc.)?

Figure D-1. RFF Regional Removal Questionnaire *(continued)*

9. What do you anticipate the impact to be of the recent decrease in the Superfund budget on
   a. FTE for removals in your region (no change/decrease/increase)
   b. Number of removals (no change/decrease/increase)
   c. Type of removals—emergency, time-critical, non-time-critical (no change/decrease/increase)
   d. Number of non-NPL removals (no change/decrease/increase)
   e. Number of NPL removals (no change/decrease/increase)
   f. Shift of NPL removal actions to the remedial program (no change/decrease/increase)

10. Given your understanding of the role of removals and the workload you have anticipated, if you could design the funding for the removals program for the next 10 years, what would make the most sense?
    a. Maintain current approach
    b. Decrease the budget
    c. Increase the budget
    d. Level fund
    e. Other approach

Appendix E

Interviews Regarding Future NPL Sites

To estimate the cost of sites added to the National Priorities List (NPL) in the future, RFF conducted two sets of interviews. We interviewed superfund program managers in 9 states and the Superfund division directors in all 10 U.S. Environmental Protection Agency (EPA) regional offices. The interviews had five main objectives:

- to identify, if possible, the specific sites the states and EPA regions expect to list on the NPL from FY 2000 through FY 2009;
- to gather information on the factors that affect NPL listing, the type of sites likely to be listed, the likelihood that future NPL sites will be expensive, and the likelihood that future NPL sites will have more or fewer Fund-lead remedial cleanups;
- to gather information on the financial and other capabilities of state cleanup programs;
- to better understand how EPA and the states perceive the role of the NPL in the overall national effort to clean up contaminated sites; and
- to determine whether states are preparing for the increase in future costs of operation and maintenance activities at sites with Fund-lead remedial actions.

State Interviews

In July 2000, RFF interviewed nine officials responsible for implementing state superfund programs. Initially, we had hoped to interview representatives of all 50 states but were unable to do so because of the restrictions placed on us by the Paperwork Reduction Act. We selected the nine states to survey based on the following three criteria:

1. *States that had listed the most sites on the NPL since FY 1996.* As shown in Table E-1, 10 states accounted for 61% of final NPL listings from October 1, 1996, through February 2, 2000.

2. *States that had the largest number of Comprehensive Environmental Response, Compensation, and Liability Information System (CERCLIS) sites with a preliminary hazard ranking system (HRS) score of 28.5 or higher.* Table E-2 lists the 10 states having the largest number of CERCLIS sites with preliminary HRS scores of at least 28.5—the threshold of eligibility for NPL listing—as of March 1998.

3. *State capability.* A more elusive but equally important criterion was state capability. We wanted to include both states with strong and well-funded state superfund programs and those with weaker and less well-funded programs. We also wanted to capture differences in regional perspective as well as differences between large and small states. To assess state capability, we relied on information from EPA regional staff about state programs as well as information from a U.S. General Accounting Office (GAO) report, *Hazardous Waste: Unaddressed Risks at Many Potential Superfund Sites* (November 1998).

Based on these three criteria, we selected the following nine states for our interviews: Massachusetts (Region 1), New Jersey (Region 2), New York (Region 2), Pennsylvania (Region 3), Florida (Region 4), Ohio (Region 5),

Table E-1. Number of Sites Listed on NPL by State, October 1996–February 2000

State	Number of sites	Cumulative percentage of total listings
New Jersey	13	13%
New York	12	25%
Texas	10	35%
Florida	5	40%
Louisiana	5	45%
California	3	48%
Georgia	3	52%
North Carolina	3	55%
Pennsylvania	3	58%
West Virginia	3	61%

Note: Based on nonfederal final NPL sites.
Source: RFF site-level dataset. FY 2000 data provided to RFF by EPA, October 2000.

Table E-2. States Containing the Most CERCLIS Sites with Hazard Ranking System Score of 28.5 or Above

State	Number of sites
Connecticut	216
Florida	195
Massachusetts	190
California	125
Illinois	112
New Jersey	112
Rhode Island	107
New York	56
Ohio	54
Tennessee	51

Source: U.S. GAO, *Hazardous Waste: Unaddressed Risks at Many Potential Superfund Sites,* November 1998, p. 68–70.

Texas (Region 6), Missouri (Region 7), and California (Region 9). All these states agreed to answer our questions on the condition that the information be kept confidential unless we explicitly obtained approval to make it public.

We emailed a draft interview guide to the participating states, EPA, and other Superfund experts in May 2000. The interview questions were revised to reflect their comments in early June and sent out to the state officials in advance of the interviews. We conducted telephone interviews in July 2000 with superfund managers in each state. Some officials provided written answers to our questions as well. A copy of the survey appears as Figure E-1.

EPA Interviews

We prepared a separate interview guide for our interviews with the Superfund division directors of the 10 EPA regional offices. Again, we sent a draft of the questions to EPA staff and a small number of other experts for their review. The final interview guide was emailed to the regions in advance of our interviews, which we conducted in May 2000. We promised to keep the information we collected confidential, except historical data, unless we specifically received permission that we could make the data public. A copy of the interview guide is included as Figure E-2.

Figure E-1. RFF State Interview Guide *(continues on next three pages)*

RESOURCES FOR THE FUTURE
STATE INTERVIEW GUIDE

Interviewee: _____ Date: _____
State: _____ Phone: _____

Current and Potential (Nonfederal) NPL Sites

1. From 1996 to the present, (*x*) sites from your state were proposed to the NPL. Why were these sites put on the NPL rather than addressed through your state cleanup program?

2. Between FY 1992 and 1995, the NPL listing rate in your state was (lower/ higher) than the rate from FY 1996 to the present. What factors in your view help to explain this change in the number of sites from your state listed on the NPL?

3. What specific sites in your state would you like to propose for NPL listing in FY 2001 and in the following three or four years? Can you identify the site by name and describe the type of site it is, whether you think it would likely be Fund- or PRP-lead, estimate the cost of cleanup ($10 million or less, $10 million–$25 million, $25 million–$50 million, over $50 million)? How long has each site been known to the state? What factors make each site a candidate for future NPL listing?

4. How does the need for states to pay 10% of remedial action costs at Fund-lead sites affect the state's view about future listings, if at all?

Sources of Supply

5. The 1998 GAO report, *Hazardous Waste: Unaddressed Risks at Many Potential Superfund Sites,* showed _____ CERCLIS sites in your state with site inspections completed and NPL decision pending to have an HRS score greater than 28.5. What is the current number of CERCLIS sites in your state with a preliminary HRS score of 28.5?

6. In your view, what percentage of these sites will be addressed by:
   _____ State enforcement program
   _____ State voluntary program
   _____ State-funded cleanup program
   _____ EPA removal program
   _____ NPL listing
   _____ EPA enforcement without NPL listing
   _____Other (please specify)

Figure E-1. RFF State Interview Guide *(continued)*

7. How does the potential universe of future sites to be cleaned up in your state compare with those currently in CERCLIS? Are you considering proposing to the NPL sites not currently in CERCLIS? Could you identify these sites?

8. Beginning in FY 1992, EPA regions implemented the Superfund Accelerated Cleanup Model. Under this policy removal actions have been used to remediate sites, and as a result, the sites often were not put on the NPL. To what extent have removals been used in your state as an alternative to NPL listing? It is our understanding that this policy is changing so that non-time-critical removal actions are being discouraged unless sites are already on the NPL. What effect might this policy change have on NPL listing rates in your state in the next few years?

9. Do you know of specific candidates for removal actions that will not be funded by EPA and thus will need to be listed on the NPL?

10. Do you foresee any major changes in the types of sites that your state will propose as candidates for listing in the next two to five years? For example, is there likely to be an increase in listing mining sites, contaminated sediment sites, Resource Conservation and Recovery Act (RCRA) sites that have gone bankrupt, eco-risk sites, and formerly used defense sites (FUDs)? Have you seen an increase in these kinds of sites needing cleanup in your state? Do you expect these types of sites to more frequently become NPL sites in the future? If so why?

11. Does your state have an active site discovery program? How much money does your state get from EPA annually for site assessment? Excluding EPA funding, what is the total state budget for site assessment and listing in FY 2000? How many state full-time equivalents (FTE) are devoted to a) discovery and b) site assessment/listing? Have resources devoted to these aspects of the program increased, decreased, or stayed the same in recent years?

12. Since FY 1996 how many sites have been discovered through your site discovery program that meet many of the criteria for NPL listing (complexity, scale, liability)? How many of these sites have been proposed to the NPL, or are likely to be proposed in the future? What has happened to the other sites?

13. To what extent has your voluntary cleanup program led to the discovery of likely future NPL sites?

Program Characteristics

14. What is the annual budget and total FTE of your state Superfund program? How much of this is funded by EPA through state superfund grants and cooperative agreements? How much is from state revenues and fees?

Figure E-1. RFF State Interview Guide *(continued)*

15. What is the annual budget and total FTE for your voluntary cleanup program? How much of this program is funded by EPA grants and cooperative agreements? How much is from state revenues or fees?

16. Does the state have the funds to clean up sites with recalcitrant or orphan PRPs under your state superfund program? If it does, can you give a few specific examples?

17. How do the enforcement authority and resources of your state superfund program compare with the enforcement authority and resources available to EPA under CERCLA? For potential sites under consideration for NPL listing or state cleanup, how does the ability to go after PRPs affect the decision whether to remediate a site under the state program or to request NPL listing?

18. How would you characterize your state program's ability (e.g., staff, enforcement authority, political will) to go after sites with:
    — recalcitrant PRPs
    — nonviable PRPs
    — many PRPs
    — expensive remedies
    — complex contamination problems
    — relocation concerns

19. What is your state's view of NPL listing—is it seen as desirable, a last resort? Why? Has the threat of NPL listing encouraged PRPs to settle with the state?

20. Does your state clean up sites that otherwise would merit NPL listing? Please identify such sites you have or are in the process of cleaning up under state authority and describe in more detail site costs, complexity, and whether or not there is a willing and viable PRP.

Operation and Maintenance (O&M)

21. How many Fund-lead NPL sites are there in your state where the state is currently responsible for O&M activities? Can you describe what kinds of O&M activities are taking place at these sites? What is the current annual budget for state O&M activities at NPL sites?

22. At how many Fund-lead NPL sites do you expect the state to become responsible for O&M over the next 5 years, next 10 years? What is your best estimate of what the future annual cost will be for these O&M activities?

Figure E-1. RFF State Interview Guide *(continued)*

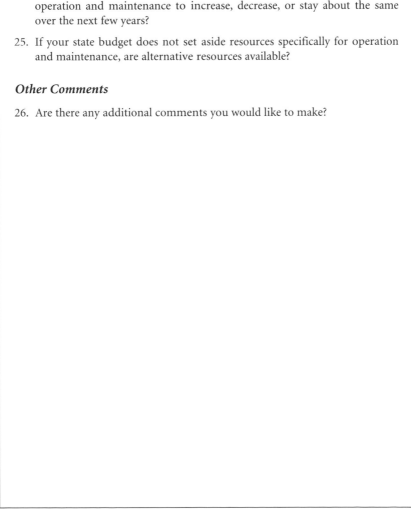

23. What is the total state budget for operation and maintenance of remedies at NPL sites this year? How many state FTE are devoted to this task? Have resources devoted to operation and maintenance increased, decreased, or stayed the same in recent years?

24. As more and more Fund-lead NPL sites enter the "construction-complete" phase of cleanup, states will be facing increasing costs for operation and maintenance at Fund-lead sites. Do you expect your total state budget for operation and maintenance to increase, decrease, or stay about the same over the next few years?

25. If your state budget does not set aside resources specifically for operation and maintenance, are alternative resources available?

Other Comments

26. Are there any additional comments you would like to make?

Figure E-2. RFF Interview Guide for EPA Regional Division Directors *(continues on next four pages)*

RESOURCES FOR THE FUTURE
INTERVIEW GUIDE: EPA REGIONAL DIVISION DIRECTORS
FUTURE NPL SITES

Interviewee: _____ Date: _____

Region: _____ Phone: _____

Recently Proposed and Final Listing NPL Sites

1. Could you tell us the following details about each site that has already been proposed for listing on the NPL in your region during FY 2000:

    Site name _____

    State location_____

    Type of site (e.g., mining, chemical manufacturing, etc.) _____

    Whether EPA or PRPs are likely to pay for removals and remedial activities or whether it is too early to tell: ____ Fund ____ PRP ____ Don't know

    How long was the region aware of the site before it was proposed? _____ years or _____ months

    Why was the state not able or willing to remediate the site?

    Could you roughly estimate the likely future cleanup costs (i.e., the cost of all future remedial actions and removals) at that site:
    ____Less than $5 million
    ____Greater than $5 million but less than $15 million
    ____Greater than $15 million but less than $25 million
    ____Greater than $25 million but less than $50 million
    ____Greater than $50 million but less than $100 million
    ____Greater than $100 million
    ____Don't know but think it is likely to be
    _____ average for this kind of site (e.g., mining, landfill, etc.)
    _____ less expensive than most
    _____ more expensive than most

2. In your view, why are states in the region putting these sites onto the NPL?

3. Between FY 1992 and 1995, the NPL listing rate in your region was *(select one)* lower *(this applies to Regions 2, 3, and 6)* / higher *(this applies to all other regions)* than the rate from 1996 to the present. What factors in your view help explain this change in your current rate of listing sites compared with the rate of listing between FY 1992 and FY 1995?

Figure E-2. RFF Interview Guide for EPA Regional Division Directors *(continued)*

4. How many of the sites that are currently proposed NPL sites in your region do you expect will make final listing? Is this likely to happen in this FY, or some time in the future?

5. In the past few years were any of the sites your region brought to the attention of headquarters—and these include both sites with submitted HRS packages and sites with the potential for an HRS score greater than 28.5—ultimately not proposed to the NPL? Why? What generally happens to these sites? Do you have a backlog of such sites from past years?

General Trends

6. During the next two to five years, do you expect the rate of NPL listings from your region to remain fairly constant, to increase, or to decrease? What kinds of sites are likely to be listed in the next five years?

7. In your region, from FY 1996 to the present, are you seeing an increase, decrease, or little change in the percent of Fund- vs. PRP-lead sites in terms of removal actions and RAs and in the percent of mega sites (that is, sites with remedies over $25 million, over $50 million, over $100 million*)?

8. In the next 5 to 10 years, what changes, if any, do you anticipate in these trends in your region (e.g., mega sites, listing rates, PRP- vs. Fund-lead sites)? What factors in your states would affect these trends?

Possible Sources of Supply to the NPL

9. How many sites do you *expect* your region will propose in the rest of FY 2000? _____ Could you tell us the following details about each of the sites:

   Site name _____

   State location _____

   Type of site (e.g., mining, chemical manufacturing, etc.) _____

   Whether EPA or PRPs are likely to pay for removals and remedial activities or whether it is too early to tell: ____ Fund ____ PRP ____ Don't know

   How long has the region been aware of the site before it is likely to be proposed? _____ years or _____ months

   Why was the state not able or willing to remediate the site?

   ---

   *Note: We later refined our definition of a mega site to be sites with total actual or expected removal and remedial costs of $50 million or more.

Figure E-2. RFF Interview Guide for EPA Regional Division Directors *(continued)*

Could you roughly estimate the likely future cleanup costs (i.e., the cost of all future remedial actions and removals) at that site:

___Less than $5 million

___Greater than $5 million but less than $15 million

___Greater than $15 million but less than $25 million

___Greater than $25 million but less than $50 million

___Greater than $50 million but less than $100 million

___Greater than $100 million

___Don't know but think it is likely to be

    ___ average for this kind of site (e.g., mining, landfill, etc.)

    ___ less expensive than most

    ___ more expensive than most

10. Does the region have a list of specific sites it would like to list on the NPL in the future? How many sites are on that list? _____

    Could you please provide us with the name of each site, what type of site it is, the location of the site, and whether it is likely to be PRP-lead (at least by the RA stage) or Fund-lead? We recognize the sensitive nature of this material and will guarantee confidentiality of site names. In our report, we will aggregate these data by site type, cost, and Fund- vs. PRP-lead.

    How many of these sites in your opinion will be listed each year for the next five years?

    What is the state's position on listing the site on the NPL?

11. How many of these sites are likely to be mega sites (that is, with removal and remedial action costs over $50 million)? Are any likely to be over $100 million? Of these, are some likely to be Fund-lead (identify number)?

12. We understand that in some regions (and in some specific states), site cleanups led by PRPs are not being put on the NPL but are being treated similarly to NPL sites. These are often referred to as NPL equivalent sites. According to information we recently got from EPA, there are ___ such sites in your region. Why are these sites not proposed for the NPL? Would these sites have been listed on the NPL 10 years ago under EPA's past listing practices?

13. How many more NPL equivalent sites are likely to be added to your list of active sites in the next 2–5 years in your region? _____

    What type of sites are they likely to be and in what states will they be located?

    Please provide information on:
    Type of site
    Fund- or PRP-lead
    Number of mega sites (and whether any of these are likely to be Fund-lead)

Figure E-2. RFF Interview Guide for EPA Regional Division Directors *(continued)*

14. One issue that has come up in our discussions with EPA and others is the question of whether there is likely to be an increase in listing of certain kinds of sites in the next few years, such as mining sites, contaminated sediment sites, RCRA sites that have gone bankrupt, eco-risk sites, and FUDs. Have you seen an increase in these kinds of sites needing cleanup? Do you expect these types of sites to become NPL sites in the future?

Site Assessment, Listing Strategies, and State Capacity

15. Our sense is that different states have different strategies for listing sites on the NPL. Could you briefly describe in your opinion the region's strategy for NPL listing, and then the strategy of each of the states in your region?

16. How would you characterize each of the states in your region in terms of:
    ___their financial capabilities and funding sources to support (a) state superfund and (b) voluntary cleanup programs
    ___the number of FTE to administer cleanup programs
    ___extent of active site discovery and assessment programs
    ___strong enforcement vs. weak enforcement
    ___likelihood of listing PRP- vs. Fund-lead sites
    ___reasons why they propose sites for NPL listing
    ___reasons why they don't want sites listed on the NPL
    ___the political dynamics of listing
    ___other issues you think are important?

17. Are there sites that "pop up" from time to time that EPA and the state really did not know about in advance, that are then proposed for NPL listing? Can you give an example or two from the past few years? Do these sites tend to be big, expensive sites or not? Is there any pattern about whether they tend to be Fund- or PRP-lead?

18. According to a report by the Association of State and Territorial Solid Waste Management Officials (ASTSWMO), during the period 1993–1997 state cleanup programs remediated some 27,000 non-NPL sites at an average cost of $600,000, a figure much lower than the average cost of an NPL cleanup. What factors explain this cost difference? In your opinion are any of the states in your region cleaning up NPL-caliber sites? Please identify specific states and provide an example of an NPL-caliber site being cleaned up by the state.

19. For each of the states in your region has the state:
    ___been paying its 10% share of remedial action costs at Fund-lead sites?
    ___and if so, has this 10% been in cash or in "in-kind" contribution?

Figure E-2. RFF Interview Guide for EPA Regional Division Directors *(continued)*

20. Does the region have an active site discovery program? What is the total regional budget for site assessment and listing in FY 2000? How many regional FTE are devoted to this task? Have resources devoted to this aspect of the program increased, decreased, or stayed the same in recent years?

21. Are there any additional comments you would like to make?

Appendix F

Building Blocks
of the RFF
Future Cost Model

In constructing our model to estimate the future cost of actions taken at current and future National Priorities List (NPL) sites, we have developed a number of building blocks:

- site types;
- unit costs of Fund-lead actions at NPL sites;
- durations of major actions and phases of cleanup at NPL sites; and
- leads of major actions at NPL sites.

Each of these building blocks is described below, as is the manner in which we adjust past expenditures into constant FY 1999 dollars and adjust FY 1999 dollars for inflation.

RFF Site Types

We code each NPL site as one of 11 site types in the RFF site-level dataset. Most of the sites were previously characterized by site type in a 1995 RFF database. Sites added to the NPL since then have been categorized by RFF using descriptive information from the U.S. Environmental Protection Agency (EPA) Superfund website. Each of the site types is described below. Definitions of the site types are based on Probst et al., *Footing the Bill for Superfund Cleanups: Who Pays and How?* (1995); the RFF database of Superfund NPL sites; and Colglazier et al., *Estimating Resource Requirements for NPL Sites* (1991).

 Chemical manufacturing sites are those at which the manufacture of chemicals is deemed the activity primarily responsible for contamination. Sites in this category include chemical plants; chemical distribution, storage, disposal, and packing centers; and pesticide and fertilizer manufacturing plants.

Oil refining sites are those at which oil refining processes are deemed the activity primarily responsible for contamination. Sites in this category include oil refineries and mud drilling operations.

Mining sites are those at which mining operations are deemed the activity primarily responsible for contamination. Sites in this category include mines and mine tailings, radium contamination, recovery plants, asbestos mills, precious metal mills, phosphate processors, and smelting facilities.

Wood preserving sites are those at which wood treating or wood preservation processes are deemed the activity primarily responsible for contamination. Sites in this category include wood preserving facilities, wood treatment facilities, and utility pole treatment facilities.

Coal gasification and other industrial sites are those at which coal gasification, the manufacture of substances or items not included in the chemical manufacturing category, or multiple industrial areas are deemed the activity primarily responsible for contamination. Sites in this category include coal gasification plants, utility plants, coal coking facilities, manufacturing plants, printing operations, tanneries, paper mills, asbestos manufacturing, foundries, and industrial parks, complexes, developments, and other properties and operations.

Recycling sites are those at which recycling is deemed the activity primarily responsible for contamination. Waste handling and disposal activities may have occurred at these sites as part of recycling activities. Sites in this category include abandoned battery recycling facilities, former drum reconditioning facilities, waste oil recycling facilities, and all other recycling facilities.

Captive waste handling and disposal sites are those where wastes are treated or disposed of, or both, and for which landfilling or waste handling and disposal are the most current uses. Captive facilities are private, generally owned by a single company, do not accept wastes on a fee-for-service basis, and accept industrial and hazardous waste for disposal. Sites in this category include captive industrial landfills and captive industrial waste management facilities.

Noncaptive waste handling and disposal sites are those where wastes are treated or disposed of, or both, and for which landfilling or waste handling and disposal are the most current use. Noncaptive facilities accept wastes for management and disposal on a fee-for-service basis, may be public or private, may or may not include a landfill, and may accept municipal, industrial, or hazardous waste. Sites in this category include municipal landfills (which may accept one or more of the above kinds of wastes), commercial landfills (which may or may not be codisposal facilities), municipal waste management facilities, and commercial waste management facilities.

Transportation sites are those at which activities related to transportation, transportation services, or both are deemed primarily responsible for contamination. Sites classified as transportation facilities include airports, cleaning operations, railroads, repair facilities, and trucking operations.

Contaminated areas are sites where the source of contamination is unknown or located off-site and therefore hard to identify. In many cases, the most likely cause of the contamination is illegal dumping or leaking underground storage tanks. However, the contamination cannot be associated with a particular facility, so the site is classified as a contaminated area rather than as a specific facility type. Contaminated areas include fields and roadsides where illegal dumping has taken place, residential areas, wells, groundwater basins, and waterways.

Miscellaneous sites are those that do not fit into the other 10 categories. Miscellaneous sites may have been formerly used as facilities in another category but have since been converted to other uses. For the purpose of analysis, sites that can be placed in more than one category because they encompass several site types are also included in the miscellaneous category. Sites in this category include agricultural facilities, dry-cleaning facilities, military facilities, multiple-operation facilities, schools and universities, and warehouses.

Unit Costs

One of the major determinants of the cost of cleaning up current and future NPL sites is the cost to EPA of implementing Fund-lead actions. Using historical expenditure data in EPA's Integrated Financial Management System (IFMS), converted into constant 1999 dollars, RFF has developed unit costs for the major actions taken at NPL sites, including remedial pipeline actions and post-construction actions. *Unless otherwise noted, all cost estimates discussed in this appendix are in constant 1999 dollars.*

Remedial Pipeline Actions (RI/FS, RD, and RA)

To calculate the average cost, which we refer to as the unit cost, for each stage of the remedial pipeline—remedial investigation/feasibility study (RI/FS), remedial design (RD), and remedial action (RA)—we use data from the RFF action-level dataset in concert with expenditure data from the RFF IFMS dataset. These unit costs are used in our projections as average costs for the completion of future Fund-lead remedial pipeline actions. Since IFMS does not track the cost of remedial pipeline actions implemented by potentially responsible parties (PRPs), we calculate unit costs based on what has historically been spent on Fund-lead actions.

We calculate the average extramural and intramural unit cost for each major remedial pipeline stage. Extramural costs are primarily the direct cost of implementing actions at each stage, such as the amount EPA pays a contractor to conduct an RI/FS, prepare an RD, or implement an RA. In IFMS, expenditures associated with contracts, interagency agreements, and cooperative agreements with states are coded as extramural costs. Intramural costs for the major pipeline actions include EPA payroll costs for staff time, such as the time a remedial project manager spends overseeing an RI/FS, as well as other associated internal expenditures, such as travel. We also include Department of Justice (DOJ) staff time as an intramural cost (even though the money is spent outside EPA).

We do not develop unit costs for PRP-lead actions, as this was not requested as part of the study; the direct costs of implementing PRP-lead RI/FSs, RDs, and RAs are paid by PRPs and do not come out of the Superfund Trust Fund. PRP-lead actions do, of course, have associated costs, both intramural (primarily EPA and DOJ staff) and extramural (e.g., oversight by contractors). These costs are captured in other categories, such as oversight, enforcement, and response support, which we estimate separately on an aggregate rather than operable unit level.

The IFMS dataset we received from EPA tracks expenditures for Fund-lead actions only at the *site* level. From these data, it is generally not possible to directly connect expenditures in IFMS to actions in EPA's Comprehensive Environmental Response, Compensation, and Liability Information System (CERCLIS) or to calculate the average cost of individual actions. For example, at a site with two operable units and two RI/FSs, all expenditures associated with the RI/FSs are simply coded as such, and it is not possible to distinguish which expenditures are for which RI/FS. In 1996, EPA began to track expenditures at the operable unit level, but staff from the Office of the Chief Financial Officer informed us that these data were reliable only for the past couple of years—not long enough, given the duration of remedial pipeline actions, for us to develop unit costs. Therefore, average costs for pipeline actions cannot be calculated at the operable unit level and have to be determined in another fashion.

We develop extramural and intramural unit costs for Fund-lead pipeline actions in the following manner. First, we sum all expenditures for a specific type of action (e.g., RI/FS) over the time period we are analyzing. The summing of dollars is straightforward. After converting all dollars into constant FY 1999 dollars (see discussion later in this appendix), we total the expenditures attributed to a specific type of action during a specific period, keeping extramural and intramural dollars separate. We use data from FY 1992 through FY 1999, since this period represents a mature Superfund program and provides us with sufficient data for analysis.

Second, we aggregate the number of Fund-lead actions (e.g., the number of Fund-lead RI/FSs) occurring during this same period, which is somewhat more complicated. We count not only the number of Fund-lead actions that start and end within the time frame, but also those actions that have only part of their duration in the time frame. This includes actions that start before the time frame and finish during it, those that start during the time frame but finish afterward, and those that start before the time frame and finish afterward.

For any action, we count only the *percentage* of that action that takes place during the period for which we tabulate costs. We use the actual start date and the actual completion date of each action, unless the action was still under way at the end of FY 1999, in which case we use the planned completion date. For example, an RI/FS that started and ended within the FY 1992–FY 1999 period counts as one RI/FS, but an RI/FS that started one year before the beginning of the period (i.e., in FY 1991) and lasted two years counts as 0.5 action, as only half the action occurred during the period we are analyzing. A similar method is used to measure partial actions for Fund-lead pipeline actions that started sometime between FY 1992 and FY 1999 but were not completed by the end of FY 1999, and for Fund-lead pipeline actions that started before FY 1992 but were still under way at the end of FY 1999. We sum the whole and partial actions to get the total number of units associated with Fund-lead costs during the period of analysis.

The final step in calculating the unit cost for a particular stage of the remedial pipeline is to divide the total expenditures by the total number of actions. For instance, to calculate the unit cost of conducting an RI/FS, we add all the expenditures (keeping extramural and intramural costs separate) attributed to RI/FSs from FY 1992 through FY 1999. We then divide this sum by the total number of whole and partial RI/FSs conducted during this period, which yields the average extramural and intramural unit cost of conducting an RI/FS. We repeat these calculations for RDs and RAs.

It is important to note that before counting the number of actions, we "collapse" all actions to ensure that we are capturing actions at the operable unit rather than the action level. That is, we count multiple remedial pipeline actions of the same type at a single operable unit as a single action. For instance, if there are two RAs at a single operable unit, we consider the entire RA for the operable unit to be the combination of these two separate RAs. As discussed below, this approach is consistent with how we define an action to estimate the average durations of different remedial pipeline stages.

We calculate average unit costs for Fund-lead RI/FSs, RDS, and RAs, controlling for the categories of sites at which the actions were conducted. The results are shown in Table F-1, broken down by pipeline stage and by all sites, mega sites, and nonmega sites. Nonmega sites are further broken down into the 11 RFF site types. The total number of units counted is also shown. This table illustrates that there are significant differences in unit costs between mega sites and nonmega sites, and among the 11 RFF site types.

To determine whether unit costs have changed over time, we calculated unit costs for two periods, FY 1992 through FY 1995, and FY 1996 through FY 1999. As before, we summed all expenditures attributed to RI/FSs, RDs, and RAs and divided by the number of actions (whole and partial) that occurred over each period. The results of these calculations are shown in Table F-2 and Table F-3.

There are significant differences in unit costs between these two periods. For all sites, mega sites, and nonmega sites, average extramural unit costs for the RI/FS and the RD phases decline from the first to the second period, whereas the average cost of the RA phase increases. However, there is variability among the RFF site types: Chemical manufacturing, recycling, captive waste handling and disposal, and contaminated areas sites show decreases in extramural unit costs for the RA phase from the first to the second period. It should also be noted that activity across pipeline phases is inconsistent, as some site types with lower RI/FS and RD unit costs after FY 1996 have higher RA unit costs, and vice versa.

In addition to this variability, there is also reason to have less confidence in the data for FY 1992–FY 1995 and for FY 1996–FY 1999 because the number of units per category has essentially been halved. Because of the extremely small sample size for each period, we decided it would be more appropriate to use the extramural unit costs derived from the expenditure data from FY 1992 through FY 1999 in our model.

Intramural unit costs also vary significantly between the two periods: They are almost zero before FY 1996, and after FY 1996 they are more than double the FY 1992–FY 1999 figures. According to EPA, the drastic increase reflects a change in the IFMS codes for staff time. As shown in Table F-4, intramural expenditures in IFMS that are coded as RI/FS, RD, and RA costs are remarkably

Table F-1. Unit Costs of Remedial Pipeline Actions (at Operable Unit Level), FY 1992–FY 1999 (1999$)

Type of site	RI/FS Extramural	Intramural	No. of units	Remedial design Extramural	Intramural	No. of units	Remedial action Extramural	Intramural	No. of units
All sites	1,324,963	27,855	283.29	1,313,691	12,036	241.79	11,010,904	30,568	240.79
Mega sites	2,479,220	40,162	63.57	3,886,939	22,109	34.57	30,364,636	62,553	51.41
Nonmega sites	991,011	24,295	219.72	884,330	10,356	207.21	5,756,692	21,884	189.38
Chemical manufacturing	711,383	18,836	20.41	826,128	24,804	20.23	5,199,644	27,775	16.18
Oil refining	552,938	483	2.20	905,203	2,265	3.00	4,510,912	32,379	2.54
Mining	2,222,175	39,603	5.14	1,172,790	15,503	8.63	10,325,396	23,212	6.45
Wood preserving	752,248	28,930	17.72	1,005,443	12,015	23.68	10,553,716	41,495	19.39
Coal gasification and other industrial	1,110,903	27,477	62.66	814,660	7,254	54.75	5,655,922	16,465	45.91
Recycling	1,163,172	18,327	18.95	1,079,953	13,224	16.44	4,418,436	12,346	14.87
Captive waste handling and disposal	439,805	8,435	10.16	776,741	19,957	6.46	3,667,780	20,533	5.82
Noncaptive waste handling and disposal	1,039,977	17,603	26.91	1,197,684	6,854	28.89	5,940,744	13,389	32.96
Transportation	559,387	9,874	10.97	447,248	4,400	2.31	10,452,720	60,914	2.98
Contaminated areas	970,140	40,111	25.99	648,690	8,048	20.12	4,058,127	29,368	16.88
Miscellaneous	1,171,268	24,567	18.62	608,855	4,642	22.71	3,198,050	18,933	25.42

Table **F-2.** Unit Costs of Remedial Pipeline Actions (at Operable Unit Level), FY 1992–FY 1995 (1999$)

Type of site	RI/FS			Remedial design			Remedial action		
	Extramural	Intramural	No. of units	Extramural	Intramural	No. of units	Extramural	Intramural	No. of units
All sites	1,440,972	51	155.37	1,329,010	0	147.68	10,176,318	11	118.58
Mega sites	2,619,711	115	33.33	3,968,924	0	18.79	26,753,232	3	26.03
Nonmega sites	1,119,027	34	122.04	944,142	0	128.89	5,513,606	14	92.55
Chemical manufacturing	866,877	42	6.82	955,077	0	12.35	5,930,137	0	7.69
Oil refining	667,601	22	1.58	640,188	0	2.77	1,146,818	0	0.94
Mining	2,336,496	0	3.78	1,922,752	0	3.28	6,742,348	0	2.37
Wood preserving	915,821	96	8.73	1,050,878	0	11.24	9,763,820	0	5.91
Coal gasification and other industrial	1,353,739	12	31.57	842,384	0	39.86	5,038,210	75	25.40
Recycling	1,143,531	96	11.92	1,045,642	0	10.61	4,664,062	0	7.71
Captive waste handling and disposal	434,112	57	8.25	1,038,212	0	1.85	9,096,189	0	1.21
Noncaptive waste handling and disposal	1,284,443	2	18.30	1,194,212	0	22.08	6,644,306	0	19.49
Transportation	635,595	0	8.30	459,229	0	1.61	2,569,826	0	0.98
Contaminated areas	1,089,550	40	8.94	1,035,031	0	6.63	6,828,810	-95	6.63
Miscellaneous	1,031,981	44	13.85	568,415	0	16.62	2,656,403	0	14.23

Table F-3. Unit Costs of Remedial Pipeline Actions (at Operable Unit Level), FY 1996–FY 1999 (1999$)

Type of site	RI/FS			Remedial design			Remedial action		
	Extramural	Intramural	No. of units	Extramural	Intramural	No. of units	Extramural	Intramural	No. of units
All sites	1,177,139	63,108	120.57	1,271,522	31,183	90.75	11,932,579	59,126	117.46
Mega sites	2,250,746	82,464	29.79	3,866,135	49,808	15.35	33,811,640	120,278	24.87
Nonmega sites	824,852	56,757	90.78	743,502	27,392	75.41	6,055,304	42,699	92.59
Chemical manufacturing	693,905	32,072	11.97	625,019	66,804	7.44	4,542,145	52,140	8.48
Oil refining	72,897	1,657	0.62	4,010,217	30,019	0.23	6,502,515	50,322	1.59
Mining	1,605,184	146,149	1.35	785,418	30,774	4.34	12,466,491	36,858	4.06
Wood preserving	773,666	71,824	6.14	961,700	22,873	12.44	11,038,039	60,290	13.31
Coal gasification and other industrial	863,784	56,489	30.29	663,397	26,419	14.87	6,447,000	36,919	19.70
Recycling	1,190,239	52,539	6.31	872,175	30,225	5.83	3,828,503	25,187	6.95
Captive waste handling and disposal	216,189	43,763	1.74	858,131	35,575	3.61	2,125,679	23,945	4.19
Noncaptive waste handling and disposal	471,165	56,926	7.95	1,193,964	29,782	6.38	6,092,374	38,894	10.90
Transportation	209,912	36,558	2.68	406,020	14,665	0.69	13,906,055	88,770	1.98
Contaminated areas	905,991	61,105	17.04	431,164	12,010	13.48	2,028,442	47,182	10.25
Miscellaneous	1,237,932	89,122	4.69	553,140	13,748	6.08	3,312,670	35,190	11.18

Table F-4. Total Intramural Expenditures for Remedial Pipeline Actions, FY 1992–FY 1999 (1999$)

Fiscal year	Total expenditures
1992	1,705
1993	5,206
1994	14,830
1995	142,649
1996	3,942,527
1997	5,322,830
1998	5,519,197
1999	7,924,071

Source: RFF IFMS dataset. Includes Fund-lead RI/FSs, remedial designs, and remedial actions only.

small from FY 1992 through FY 1995, and then increase nearly 30-fold between FY 1995 and FY 1996.

Because of this anomaly, we decided when constructing our model not to use intramural expenditure data before FY 1996 and instead to use intramural unit costs calculated from expenditure data from FY 1996 through FY 1999. Also, because of the small number of units in certain RFF site types, as discussed above, we decided not to assign intramural costs by site type. Instead, we vary intramural unit costs only by whether the action occurred at a mega or nonmega site. Although we would have preferred to use the same time frame for our estimates of extramural and intramural unit costs, problems with the data would not support such an approach.

LTRA and Five-Year Reviews

We are unable to compute unit costs for long-term response actions (LTRAs) and five-year reviews using the same methodology as for remedial pipeline actions. The main obstacle is a lack of data in both CERCLIS and IFMS. LTRAs are significantly underrepresented in CERCLIS, and the IFMS codes for LTRAs and five-year reviews have not been consistently used by the Agency since they were created in 1996.

For LTRAs, as an alternative, we estimate annual costs based on analyses of data from the RFF remedial action cost database (see Appendix D) and a 1999

EPA report.[1] The average annual LTRA cost, calculated from all remedial action contracts with LTRAs represented in the RFF remedial action cost database, is approximately $760,000. The average annual LTRA costs at mega sites and nonmega sites are $2 million and $450,000, respectively. We do not find significant variation among the average annual costs of LTRAs at non-mega sites across site types.

The 1999 EPA report contains cost information on the operation of groundwater treatment systems, which are used for purposes analogous to LTRA. The average annual cost of operating these systems at nonmega sites and mega sites is $355,000 and $3.3 million, respectively. Based on this information, our best judgement is that the average LTRA costs about $400,000 annually for an operable unit at a nonmega site, and $2 million annually for an operable unit at a mega site.

We are unable to calculate an average cost for a five-year review using expenditure data from the RFF IFMS dataset. The IFMS code for five-year reviews has been used only since 1996, and as with LTRAs, there are insufficient data to estimate the cost of conducting a five-year review. We instead base our estimate of the cost of carrying out a five-year review—$25,000 for a nonmega site—on an estimate in a 1999 EPA Office of the Inspector General report.[2] Since mega sites are generally larger and more complex than non-mega sites, we assume that a five-year review at a mega site costs twice as much, or $50,000.

Durations

We develop average durations for each of the main phases of the remedial pipeline (RI/FS, RD, and RA), as well as for the periods between phases, using the RFF action-level dataset. As noted in Appendix C, the RFF action-level dataset has been created from CERCLIS data (including EPA planning data).

The general method for calculating the duration of an action is to count the days between the start date and the completion date. For actions under way at the end of FY 1999 that do not have completion dates, we use EPA's planned completion dates. This approach differs significantly from that of EPA, which calculates durations based solely on actions that have started and been completed within the time frame being analyzed. Thus, EPA's average durations provided to RFF in May 2000 (presented in Table F-5), which represent average durations for actions completed from FY 1993 through FY 1999, exclude 191 RI/FSs, 225 RDs, and 464 RAs still under way at the end of

FY 1999.* EPA also calculates average durations based on individual actions at an operable unit, rather than when a phase in the pipeline starts and ends for the specific operable unit. There are often multiple actions of the same type at one operable unit (for example, multiple RI/FSs). We "collapse" these actions to estimate average durations because this approach better captures how long a phase of the pipeline takes. That is, we give an action the start date of the earliest action at that operable unit and the completion date of the last action. Therefore, if one action was still under way at the end of FY 1999, we deem the entire action at that operable unit under way as well.

In evaluating the data in the RFF action-level dataset, we found that there were 123 RAs under way at the end of FY 1999 that had expected durations of more than 10 years, which seemed unusual. We worked with EPA to verify the status of these "long RAs" and, based on the information that the Agency provided, made the following corrections and adjustments to the RFF action-level dataset:

- 30 RAs were complete and the operable unit had moved into LTRA; this information had not been entered into CERLCIS, so we assign the RAs their proper completion dates and create the missing LTRA actions;
- 46 RAs were scheduled to transition into LTRA but had incorrect planned completion dates because of a change in EPA coding practices in CERCLIS; EPA provided new planned completion dates for these RAs, and a future LTRA was created and given a start date equivalent to the RA completion date;
- 7 RAs were complete but were miscoded as still being under way;
- 4 RAs were given revised completion dates based on EPA explanations; and
- 36 RAs were determined to have the correct planned completion date; most of these RAs include such remedies as monitored natural attenuation, bioremediation, and soil vapor extraction, each of which can require decades of operation to meet cleanup goals.

After making these changes, we calculated average durations, using data from actions started in FY 1992 and later at the 1,245 sites listed as final or deleted on the NPL at the end of FY 1999. For the RI/FS duration, we use the time from the start of the RI/FS to the date the record of decision (ROD) is signed (the signing of the ROD is the formal conclusion of the RI/FS process). We measure the RI/FS-to-RD duration as the time between the ROD signing

---

*The figures are based on "uncollapsed" actions at NPL final and deleted sites and count only RIs and FSs conducted concurrently.

date and the start of the RD. Also, because an RA typically begins before its associated RD is completed, the average time between the RD stage and the RA stage is negative. Rather than use a negative duration, we add the negative value to the duration of the RD stage.

Including long RAs (those lasting 10 years or longer) in the RA duration results in a bimodal distribution, with a large cluster of RAs relatively close to 3 years in duration and a small cluster of RAs closer to 20 years. Overall, 4% of RAs that started in FY 1992 or later have expected durations of 10 years or more. These long RAs are not random but mostly are cases in which bioremediation or soil vapor extraction is part of the remedy. To account for these long RAs when making our projections for actions at current and future NPL sites, we calculate two sets of RA durations, one for RAs with durations of less than 10 years, and one for RAs with durations of 10 years or longer—an arbitrarily chosen cutoff point. RA durations are shown in Table F-5. Although the average duration for all RAs is 3.98 years, 96% of RAs have an average duration of 3.29 years, and the remaining 4%—the long RAs—have an average duration of 20.44 years.

We analyzed durations in a number of ways. We considered whether an action has been conducted (or is being conducted) at a mega or nonmega site, whether an action has been conducted (or is being conducted) at a teenager or nonteenager site, and whether an action is Fund-lead or PRP-lead. Table F-5 shows the average durations for the RI/FS, RI/FS-to-RD, RD, and RA stages, grouped by these attributes. The table also shows the sum of durations from the start of an RI/FS to the end of an RA for all three RA measurements. It is important to emphasize that this sum of durations is for a single operable unit, not for a site as a whole.

We also analyzed historical data from FY 1992 through FY 1999 to estimate the average time from the date a site is proposed to the NPL to the start of its first RI/FS. There is a significant difference depending on the lead of the action. We calculated the average time to the first RI/FS when it is Fund-lead to be about 0.53 years, whereas the average time to the first RI/FS when it is PRP-lead is about 1.21 years. (In these calculations we used only data from nonteenager sites, as the first RI/FSs were conducted for teenager sites before our cutoff of FY 1992.) As discussed in Chapter 3, the difference is most likely due to the extra time it takes to identify and negotiate a settlement with PRPs before the RI/FS starts.

We also calculated the average time from the date a site is listed as final on the NPL to the first RI/FS. As discussed in Appendix G, it is necessary to calculate this duration because our model for the future NPL estimates the cost

Table F-5. Durations of Remedial Pipeline Phases (Years)

Type of site	RI/FS	RI/FS to remedial design	Remedial design	Remedial action			Total duration		
				All	Less than 10 years (96%)	Long (4%)	Using all remedial action	Using remedial action less than 10 years only	Using long RA only
EPA									
All sites	2.60	1.00	1.70	1.80	—	—	7.10	—	—
RFF									
All sites	4.14	0.76	2.25	3.98	3.29	20.44	11.12	10.44	27.59
Nonmega	3.91	0.78	2.27	3.93	3.19	20.99	10.89	10.16	27.95
Mega	4.99	0.64	2.11	4.25	3.84	16.65	11.98	11.58	24.39
Fund-lead	3.81	0.28	2.32	4.03	3.81	17.84	10.44	10.22	24.25
PRP-lead	5.04	0.97	2.18	3.89	3.08	21.67	12.08	11.27	29.86
Nonmega, nonteenager	3.64	0.68	2.10	3.77	3.01	21.88	10.19	9.43	28.30
Nonmega, teenager	4.88	1.03	2.65	4.45	3.80	18.30	13.01	12.36	26.86
Mega, nonteenager	5.03	0.29	2.03	3.83	3.21	31.21	11.18	10.56	38.56
Mega, teenager	4.96	0.81	2.08	4.48	4.19	11.79	12.31	12.03	19.63

Note: Totals do not include the duration between NPL proposal date and start of first RI/FS. Totals may not add exactly because of rounding.

Source: EPA durations provided to RFF by EPA, May 2000, and are based on completed actions (FY 1993–FY 1999).

for each operable unit beginning with the year the sites are listed as final, not when they are proposed. Because of the increased tendency to begin RI/FSs before a site is listed as final, the average duration from a site's NPL final listing date to the start of its first RI/FS is negative.

As discussed in Chapter 4, we are unable to calculate the duration of LTRAs using the same methodology as for other remedial pipeline actions. The National Contingency Plan limits LTRA to 10 years, at the end of which, if continued groundwater or surface water restoration activities are necessary, they become the responsibility of the state in which the site is located and cannot be paid for with Trust Fund monies. Although EPA may conduct LTRA for 10 years, using data from the RFF remedial action cost database, we find that for the 35 LTRAs to be conducted at nonmega sites for which we have complete information, the average planned duration is approximately nine years. In our model, we therefore assume that LTRAs at nonmega NPL sites (both current and future) will last nine years. For mega sites on the current NPL, we use site-specific LTRA duration information. We assume that the average duration for LTRAs at future mega sites will be 10 years. Where EPA is overseeing PRP-lead groundwater or surface water restoration activities (referred to as PRP LR in CERCLIS), we use an average duration of 30 years, since there is no time limit and PRP LR is not completed until the remedial action goals set forth in the operable unit's ROD have been achieved, which generally takes decades.

In our model, we use EPA planning data where available to project when future actions will be undertaken. For each operable unit, the planning data provided by the Agency include the planned completion date for all actions under way and the planned start date for the next action at each operable unit. For those actions for which we do not have EPA planning data, we use the average durations, as calculated above, to predict when actions under way will be completed and when future actions will begin. For actions at nonmega sites on the current NPL, we use two sets of durations—one for teenager sites and one for nonteenager sites—because on an aggregate level (start of RI/FS through completion of RA), the difference between teenager and nonteenager sites is more significant than for the other categories of sites. These durations are shown in Table F-6. For actions at mega sites on the current NPL, we use site-specific data from the RFF mega site survey. For actions at sites added to the NPL from FY 2000 through FY 2009, we use average mega site and nonmega site durations calculated from actions conducted (or being conducted) only at nonteenager sites, as shown in Table F-7. We use different durations for actions to be conducted at future NPL sites than for actions to be con-

Table F-6. Durations of Remedial Pipeline Phases at Nonmega Sites on Current NPL

Pipeline phase	Teenager sites			Nonteenager sites		
	Calculated (years)	Rounded (years)	Rounded (quarters)	Calculated (years)	Rounded (years)	Rounded (quarters)
NPL proposal date to first RI/FS (Fund-lead)[a]	—	—	—	0.53	0.50	2
NPL proposal date to first RI/FS (PRP-lead)[a]	—	—	—	1.21	1.25	5
RI/FS	4.88	5.00	20	3.64	3.75	15
RI/FS completion to remedial design start	1.03	1.00	4	0.68	0.75	3
Remedial design	2.65	2.75	11	2.10	2.00	8
Remedial action	3.80	3.75	15	3.01	3.00	12
Long RA	20.44	20.50	82	20.44	20.50	82
LTRA	9.00	9.00	36	9.00	9.00	36
PRP LR	30.00	30.00	120	30.00	30.00	120

[a] Durations for this phase were not calculated for teenager sites for reasons detailed earlier.

ducted at current NPL sites, because we assume that sites added to the NPL from FY 2000 through FY 2009 will not be teenager sites. To address the issue of long RAs, we also use two durations for RAs. All durations used in the model are rounded to the nearest fiscal quarter.

Action Leads

Definition of Fund-Lead and PRP-Lead

Although fundamentally EPA characterizes remedial pipeline actions as either Fund-lead or PRP-lead, the coding system used by the Agency is actually more complex. For the purposes of our research and analyses, we slightly modify EPA's categories. Table F-8 shows EPA's original categories and our modifications to them.

Table F-7. Durations of Remedial Pipeline Phases at Future NPL Sites

	Mega sites			Nonmega sites		
Pipeline phase	Calculated (years)	Rounded (years)	Rounded (quarters)	Calculated (years)	Rounded (years)	Rounded (quarters)
NPL final date to first RI/FS	0.13	0.25	1	–0.19	0.00	0
RI/FS	5.03	5.00	20	3.64	3.50	14
RI/FS completion to remedial design start	0.29	0.25	1	0.68	0.75	3
Remedial design	2.03	2.00	8	2.10	2.00	8
Remedial action	3.21	3.25	13	3.01	3.00	12
Long RA	20.44	20.50	82	20.44	20.50	82
LTRA	10.00	10.00	40	9.00	9.00	36
PRP LR	30.00	30.00	120	30.00	30.00	120

Note: To incorporate the negative average duration at nonmega sites for the phase "NPL final date to first RI/FS," this value is added to the average duration of the RI/FS phase prior to rounding.

Historical Lead Data

At the request of RFF, in October 2000, EPA's Office of Enforcement and Compliance Assurance (OECA) provided historical data on the leads for each major remedial pipeline stage. The data indicated the percentage of Fund-lead versus PRP-lead actions—RI/FSs, RDs, and RAs—for the following three periods:

1. from FY 1980 through FY 1986 (from the inception of the Superfund program until the Superfund Amendments and Reauthorization Act (SARA);
2. from FY 1987 through FY 1990 (the years after SARA but before implementation of the enforcement-first policy); and
3. from FY 1991 through FY 1999 (the years since implementation of the enforcement-first policy).

OECA organized the leads for the different actions by the 11 RFF site types and by mega and nonmega sites. OECA also broke down the data to the annual level for FY 1991 through FY 1999.

Table F-9 shows these historical lead data for all sites, mega sites, and nonmega sites by type of action for the three periods: FY 1991 through FY 1995,

Table F-8. EPA and RFF Definitions of Action Leads

Fund	PRP	Other/federal facility
EPA		
• EPA Fund financed (F) • State Fund financed (S) • Federal enforcement (FE) • State enforcement (SE) • Tribal Fund financed (TR) • EPA in-house (EP) • Coast Guard (CG)	• Responsible party (RP) • PRP response under state (PS) • Mixed funding (MR)	• State no Fund money (SN) • PRP-lead under state (SR) • Federal facility at nonfederal site (FF)
RFF		
• EPA Fund financed (F) • State Fund financed (S) • Federal enforcement (FE) • State enforcement (SE) • Tribal Fund financed (TR) • EPA in-house (EP) • Coast Guard (CG)	• Responsible party (RP) • PRP response under state (PS) • Mixed funding (MR) • State no Fund money (SN) • PRP-lead under state (SR) • Federal facility at nonfederal site (FF)	

Note: CERCLIS code is in parentheses. CERCLIS codes that are not used for removal actions, RI/FSs, remedial designs, or remedial actions are excluded.

FY 1996 through FY 1999, and FY 1991 through FY 1999. Table F-10 shows these data by site type for FY 1991 through FY 1999.

In our model of the current NPL, we use site-specific information on the leads of all future actions at mega sites collected as part of the RFF mega site survey. For nonmega sites, we use lead information from EPA planning data when available. For actions at nonmega sites for which we do not have lead information as part of the planning data and for future actions not included in the planning data, we use the historical action lead data from FY 1991 through FY 1999, by site type, as shown in Table F-10. For actions at future NPL sites, since we are not able to predict the distribution of site types, it is not possible to assign leads of actions by site type. Instead, we base our future lead projections on recent history, using the overall proportion of Fund-lead to PRP-lead actions from FY 1996 through FY 1999, for mega and nonmega sites, as shown in Table F-9. We use these historical proportions (rounded) for the low-case scenario of the future NPL and adjust them for the base and high cases, based on our interviews with officials in the 10 EPA regions and 9 state agencies.

Table F-9. Remedial Pipeline Action Lead History by Category of Site

	RI/FS				Remedial design				Remedial action			
	Fund-lead		PRP-lead		Fund-lead		PRP-lead		Fund-lead		PRP-lead	
	Number	%	Number	%	Number	%	Number	%	Number	%	Number	%
All sites												
FY 1991–FY 1995	143	45%	172	55%	206	27%	551	73%	182	27%	481	73%
FY 1996–FY 1999	100	76%	32	24%	97	29%	237	71%	125	26%	347	74%
FY 1991–FY 1999	243	54%	204	46%	303	28%	788	72%	307	27%	828	73%
Nonmega sites												
FY 1991–FY 1995	105	45%	130	55%	176	28%	449	72%	142	26%	401	74%
FY 1996–FY 1999	77	75%	26	25%	76	28%	198	72%	103	27%	283	73%
FY 1991–FY 1999	182	54%	156	46%	252	28%	647	72%	245	26%	684	74%
Mega sites												
FY 1991–FY 1995	38	48%	42	53%	30	23%	102	77%	40	33%	80	67%
FY 1996–FY 1999	23	79%	6	21%	21	35%	39	65%	22	26%	64	74%
FY 1991–FY 1999	61	56%	48	44%	51	27%	141	73%	62	30%	144	70%

Note: Percentages may not add to 100% because of rounding.
Source: Data provided to RFF by EPA, October 2000.

Table F-10. Remedial Pipeline Action Lead History at Nonmega Sites by Site Type, FY 1991–FY 1999

| | RI/FS | | | | Remedial design | | | | Remedial action | | | |
| | Fund-lead | | PRP-lead | | Fund-lead | | PRP-lead | | Fund-lead | | PRP-lead | |
	Number	%	Number	%	Number	%	Number	%	Number	%	Number	%
Chemical manufacturing	21	45%	26	55%	26	32%	55	68%	26	35%	48	65%
Oil refining	2	67%	1	33%	3	25%	9	75%	3	21%	11	79%
Mining	2	40%	3	60%	20	83%	4	17%	24	67%	12	33%
Wood preserving	21	78%	6	22%	38	57%	29	43%	29	44%	37	56%
Coal gasification and other industrial	52	55%	42	45%	61	31%	133	69%	59	29%	144	71%
Recycling	11	58%	8	42%	14	18%	65	82%	16	19%	70	81%
Captive waste handling and disposal	8	38%	13	62%	8	14%	51	86%	7	11%	59	89%
Noncaptive waste handling and disposal	14	25%	42	75%	28	12%	211	88%	28	11%	217	89%
Transportation	7	88%	1	13%	2	11%	16	89%	4	29%	10	71%
Contaminated areas	32	73%	12	27%	26	36%	47	64%	23	37%	39	63%
Miscellaneous	12	86%	2	14%	26	49%	27	51%	26	41%	37	59%

Note: Percentages may not add to 100% because of rounding.

Source: Data provided to RFF by EPA, October 2000.

Adjusting for Inflation

One key component in our model is the use of unit (or average) costs for the phases of the remedial pipeline. As noted earlier, these costs are derived from historical expenditure data. To ensure that our unit costs represent the true costs of these actions for future years, we convert all expenditures made prior to FY 1999 into "constant 1999 dollars" (1999$). In other words, we adjust expenditures in earlier years to what they would have been in 1999, if future inflation had been taken into account.

We adjust past expenditures using the U.S. Department of Commerce Bureau of Economic Analysis (BEA) measurements of the U.S. gross domestic product (GDP), as reported in BEA's National Income and Product Accounts tables in current and 1996 dollars, by calendar year. We use these data to calculate a factor to convert expenditures in each year from current dollars into 1999 dollars, as shown in Table F-11.

To account for future inflation, we adjust our estimates made for FY 2000 through FY 2009. To do this, we use the GDP price index forecasted and projected by the Congressional Budget Office, included in *The Budget and Economic Outlook* in January 2001. We use the actual GDP price index for FY 2000. We then compute a factor to convert 1999 dollars into inflation-adjusted dollars. The GDP price index projections and conversion factor are shown in Table F-12.

Table F-11. Conversion from Current Dollars to 1999 Dollars

| Calendar year | Gross domestic product | | Conversion factor: current dollars to 1999 dollars |
	Current dollars (billions)	1996 dollars (billions)	
1980	2,795.6	4,900.9	1.8339
1981	3,131.3	5,021.0	1.6774
1982	3,259.2	4,919.3	1.5789
1983	3,534.9	5,132.3	1.5188
1984	3,932.7	5,505.2	1.4644
1985	4,213.0	5,717.1	1.4196
1986	4,452.9	5,912.4	1.3890
1987	4,742.5	6,113.3	1.3485
1988	5,108.3	6,368.4	1.3041
1989	5,489.1	6,591.8	1.2562
1990	5,803.2	6,707.9	1.2092
1991	5,986.2	6,676.4	1.1667
1992	6,318.9	6,880.0	1.1390
1993	6,642.3	7,062.6	1.1123
1994	7,054.3	7,347.7	1.0896
1995	7,400.5	7,543.8	1.0664
1996	7,813.2	7,813.2	1.0461
1997	8,300.8	8,144.8	1.0264
1998	8,759.9	8,495.7	1.0145
1999	9,256.1	8,848.2	1.0000

Source: Bureau of Economic Analysis, National Income and Product Accounts, May 2000.

Table F-12. Conversion from 1999 Dollars to Inflation-Adjusted Dollars

	Gross domestic product price index		Conversion factor:
Fiscal year	Percentage change over previous year	Multiplier over previous year	1999 dollars to inflation-adjusted dollars
2000	1.9%	1.0190	1.0190
2001	2.3%	1.0230	1.0424
2002	2.2%	1.0220	1.0654
2003	2.1%	1.0210	1.0877
2004	1.9%	1.0190	1.1084
2005	1.9%	1.0190	1.1295
2006	1.9%	1.0190	1.1509
2007	1.9%	1.0190	1.1728
2008	1.9%	1.0190	1.1951
2009	1.9%	1.0190	1.2178

Source: Congressional Budget Office, *The Budget and Economic Outlook: Fiscal Years 2002–2011,* January 2001, p. 135.

The RFF
Future Cost Model

To estimate the future cost to the U.S. Environmental Protection Agency (EPA) of the Superfund program, we group expenses incurred by the Agency into two broad categories: those associated with Fund-lead cleanup actions at sites on the National Priorities List (NPL) and those associated with other parts of the program. The reason for making estimates in this manner is driven, in large part, by the quality and availability of data, but also by our interest in creating a model that is transparent and grounded in the real world.

This appendix explains step-by-step how we develop the model to estimate the cost to EPA of the Superfund program from FY 2000 through FY 2009. For costs associated with Fund-lead cleanup actions at current NPL sites (listed as final or deleted as of the end of FY 1999), we make our estimates at the operable unit level, and for future NPL sites (listed as final from FY 2000 through FY 2009), at the site level. We estimate five-year reviews at the site level, and all other future costs at the function level (e.g., enforcement, oversight, staff, management, support, administration) or at the programmatic or agency level (e.g., removal program, site assessment, U.S. Coast Guard). Table G-1 summarizes how we group the various components of the Superfund program and indicates each component's FY 1999 expenditures. In addition, we indicate the approach we take to estimate the future cost of each component.

The first section of this appendix describes our methodology for estimating the future expense to EPA of implementing Fund-lead cleanup actions at sites on the current NPL. Included is an explanation of our methodology for projecting when sites will reach construction completion and when five-year reviews will be necessary. The second section describes the approach we use to estimate the future cost to EPA of implementing Fund-lead cleanup actions at sites added to the NPL from FY 2000 through FY 2009—the future NPL. The final

Table G-1. FY 1999 Superfund Expenditures and RFF Estimation Approach

Program element	FY 1999 expenditures	Method of cost projection
Removal program		
NPL final and deleted sites	80,309,931	Hold constant
Non-NPL sites	140,065,969	Hold constant
Oversight of PRP-lead actions	7,609,856	Hold constant
Department of Justice enforcement	4,875,157	Hold constant
EPA enforcement	9,791,119	Hold constant
Response support	28,083,223	Hold constant
Community and state involvement	5,510,740	Hold constant
Staff, management, and support	41,600,000	Hold constant
Subtotal	**317,845,995**	
Site assessment	**57,802,949**	Hold constant
Remedial program		
NPL final and deleted sites		
Fund-lead RI/FS, remedial design, remedial action	465,298,124	Model (current NPL and future NPL)
Long-term response action	12,844,313	Model (current NPL and future NPL)
Oversight of PRP-lead actions	33,197,697	Adjusted PRP actions scaling factor
Department of Justice enforcement		
Cost recovery	20,225,171	Department of Justice cost recovery scaling factor
Other	5,056,293	NPL activity scaling factor
Subtotal	*25,281,464*	
EPA enforcement		
Cost recovery	7,633,912	EPA cost recovery scaling factor
Maximizing PRP involvement	22,901,737	NPL activity scaling factor
Subtotal	*30,535,649*	
Response support	15,326,026	Adjusted all-actions scaling factor
Community and state involvement	9,919,339	Adjusted all-actions scaling factor

continued on next page

Table G-1. *continued*

Program element	FY 1999 expenditures	Method of cost projection
Five-year reviews	513,354	Model (current NPL and future NPL)
Subtotal	*592,915,966*	
Non-NPL sites		
Fund-lead RI/FS, remedial design, remedial action	12,333,713	Hold constant
Oversight of PRP-lead actions	2,216,128	Hold constant
Department of Justice enforcement	395,797	Hold constant
EPA enforcement	2,295,157	Hold constant
Response support	11,467,080	Hold constant
Community and state involvement	691,781	Hold constant
Subtotal	*29,399,656*	
Subtotal	**622,315,622**	

Program staff, management, and support

Program element	FY 1999 expenditures	Method of cost projection
Remedial program policy, management, and staff	113,900,000	Adjusted all-actions scaling factor
Enforcement program policy, management, and staff		
NPL related	75,593,106	NPL activity scaling factor
Non-NPL related	23,506,894	Hold constant
Subtotal	*99,100,000*	
Response action management	37,500,000	Adjusted all-actions scaling factor
State programs and grants	31,400,000	Hold constant
Laboratory support	16,600,000	Adjusted all-actions scaling factor
Office of Solid Waste and Emergency Response front office	4,000,000	Hold constant
Office of Enforcement and Compliance Assurance front office and communications	2,300,000	Hold constant
National Enforcement Investigations Center	13,700,000	Hold constant
Office of Research and Development	4,000,000	Hold constant
Office of General Counsel	3,100,000	Hold constant
Subtotal	**325,600,000**	

continued on next page

Table G-1. *continued*

Program element	FY 1999 expenditures	Method of cost projection
Program Administration		
Rent, facilities operation and maintenance, and equipment and supplies	61,200,000	Total dollars scaling factor
Office of Administration and Resources Management		
Activities and regional staff	32,300,000	Total dollars scaling factor
Information systems and services	2,400,000	Hold constant
Subtotal	*34,700,000*	
Office of the Chief Financial Officer activities	18,200,000	Total dollars scaling factor
Subtotal	**114,100,000**	
Other programs and agencies		
Technology Innovation Office	8,100,000	Hold constant
Chemical Emergency Preparedness and Prevention Office	7,600,000	Hold constant
National Institute for Environmental Health Sciences	60,000,000	Hold constant
United States Coast Guard	4,800,000	Hold constant
Office of the Inspector General	20,000,000	Hold constant
Subtotal	**100,500,000**	
Total Superfund Expenditures	*1,538,164,566*	

section of the appendix discusses our methodology for estimating the costs of the other components of the Superfund program, which we do either by scaling these costs to indicators of the level of activity in the Superfund program or by holding them constant at their FY 1999 expenditure level.

Cost of Fund-Lead Actions at Current NPL Sites

Nearly a third of FY 1999 Superfund expenditures, or $478.1 million of $1.54 billion, was spent for Fund-lead cleanup actions—remedial investigation/feasibility studies (RI/FSs), remedial designs (RDs), remedial actions (RAs), and long-term response actions (LTRAs)—at NPL sites. These costs are, and will continue to be, the most significant single expense EPA incurs in its implementation of the Superfund program. The first component of our model focuses on estimating these costs. Our estimates of Fund-lead actions are not fully loaded, meaning they do not represent the total resources required to implement Fund-lead cleanup activities but rather they represent only those associated with direct cleanup. Management and contract administration costs, for example, are not included.

In Chapter 7 we present estimates for the cost of Fund-lead actions at current NPL sites under three scenarios. Using the same model, we change specific assumptions (e.g., the cost of remedial actions) to generate two alternatives to the base-case scenario. The model described below pertains to all three scenarios. For the purpose of estimating these costs, we separate the 1,245 sites listed as final or deleted on the NPL at the end of FY 1999 into two categories: mega sites, designated by EPA as having actual or expected total removal and remedial action costs of $50 million or more, and nonmega sites, with actual or expected total removal and remedial action costs of less than $50 million. For mega sites, we base our estimates primarily on data collected as part of the RFF mega site survey, described in Appendix D. For nonmega sites, we estimate the number, duration, and cost of ongoing and future Fund-lead actions based on our analyses of historical data from EPA's databases, the Comprehensive Environmental Response, Compensation, and Liability Information System (CERCLIS) and the Integrated Financial Management System (IFMS). We describe the approach we use for both categories of sites below.

Mega Sites on the Current NPL

As described in Appendix D, we collected site-specific data for each of the 112 mega sites on the current NPL identified by EPA. Because we have site-specific information for all future actions at mega sites on the current NPL (we define the current NPL as all final and deleted sites on the NPL as of the end of FY 1999), we assume that there will not be "no further action" (NFA) records of decision (RODs) or increases in the number of operable units ("OU growth") at any of these sites. (These factors are discussed below in detail in reference to nonmega sites.) Table G-2 reports the total number of ongoing and future actions at the 112 mega sites at the end of FY 1999 as indicated by the remedial project managers (RPMs), as well as the number of actions already completed at these sites based on information from the RFF action-level dataset.

Table G-2. Summary of Remedial Pipeline Actions, by Type, at Mega Sites on the Current NPL, End of FY 1999

	RI/FS		Remedial design		Remedial action		LTRA	
	Fund-lead	PRP-lead	Fund-lead	PRP-lead	Fund-lead	PRP-lead	Fund-lead	PRP-lead
Actions completed[a]	129	71	78	118	52	67	4	0
Actions under way	41	53	20	40	34	79	8	9
Actions not yet started	9	13	43	97	49	120	20	51
Total	**179**	**137**	**141**	**255**	**135**	**266**	**32**	**60**

[a]Based on RFF action-level dataset. PRP-lead LTRA is also referred to as PRP LR.

We total the RPMs' estimates of future extramural costs for Fund-lead actions across all mega sites to obtain the total extramural cost of Fund-lead actions by fiscal year. (Also included are costs associated with actions funded by potentially responsible parties, PRPs, because at a small number of sites, some PRP-lead actions are partially paid for with Trust Fund monies.) As noted in Appendix D, we did not request estimates of intramural costs (primarily staff time associated with Fund-lead actions) as part of the survey; instead, we use intramural unit costs calculated from IFMS expenditure data. As detailed in Appendix F, using IFMS expenditure data from FY 1996 through FY 1999, the average intramural cost at the operable unit level associated with each RI/FS, RD, and RA conducted at a mega site is $82,464, $49,808, and $120,278, respectively. (Intramural costs associated with LTRAs are captured elsewhere in our model.) For each action, we distribute the appropriate extramural unit cost uniformly over every year in which these costs are incurred. We total these site-specific estimates of future intramural costs across all mega sites to obtain total intramural cost by fiscal year. Lastly, we sum these intramural costs with extramural costs by fiscal year for FY 2000 through FY 2009, which yields our total estimates for the future cost of Fund-lead cleanup actions at those 112 mega sites on the NPL at the end of FY 1999.

Nonmega Sites on the Current NPL

We did not collect site-specific estimates of future costs for the 1,133 non-mega sites listed as final or deleted on the current NPL at the end of FY 1999. Instead, we first use the RFF action-level dataset (described in Appendix C) to determine the cleanup status of the 1,724 operable units at the 1,133 nonmega sites. We then estimate the number, duration, and cost of all ongoing and future Fund-lead cleanup actions, based on our analyses of CERCLIS and IFMS expenditure data, discussed in Chapters 3 and 4 and in Appendix F. Our approach is detailed below.

Baseline Dataset

The first step in estimating the anticipated cost of Fund-lead actions at non-mega sites is to determine each operable unit's status within the remedial pipeline at the end of FY 1999. We assign each operable unit to one of four categories: (1) those with a pipeline action under way, (2) those between pipeline phases that have a past completed action but do not have a subsequent action under way, (3) those without a cleanup action completed or under way, and (4) those at the end of the pipeline or for which no further remedial pipeline action is planned.

For each operable unit, we identify a last action, by which we mean the status of each operable unit as of September 30, 1999, which we use as the action from which to project the operable unit through the remainder of the remedial pipeline. Ideally, the last action at an operable unit would be simply the most recent stage of the pipeline. However, as discussed in Chapter 3, more than one action of a given type may be undertaken at a single operable unit, or remedial activity may retreat to an earlier stage after the discovery of new information. For example, at some sites an operable unit might have a completed RI/FS, RD, and RA before it is discovered that it needs a second RI/FS. In such a case, even though the RA is the last action in the remedial pipeline, the second RI/FS becomes the last action so that we can capture the future costs associated with the RD and RA that will follow. To account for the different possibilities, we determine the last action for each operable unit according to the following decision rules:

- If an operable unit has an action under way at the end of FY 1999, that is the last action. (This is the case at 791 operable units.)
- If an operable unit has no action under way at the end of FY 1999 but does have past completed actions, the action with the latest completion date is identified as the last action. (This is the case at 706 operable units.)
- If an operable unit has a ROD calling for no further action, we do not project it through the pipeline, and we do not assign it a last action. (This is the case at 122 operable units.)
- If an operable unit has multiple actions under way, we sort them by their planned completion dates and select the latest one. As a check, we also

consider these operable units on a case-by-case basis. If an operable unit has multiple actions under way, no matter what their stages, we include the likely costs of completing all the under way actions. (This is the case at 89 operable units.)

- If an operable unit has no actions completed or under way, we do not assign it a last action. We assume it is at the beginning of the remedial pipeline and will go through all phases of the remedial pipeline, unless the site has already been deemed construction-complete. (This is the case at 16 operable units.)

In the course of determining the last action for each operable unit, it was evident that we had to clean up some other parts of the baseline dataset because of missing or incorrect CERCLIS data, since these data form the basis of the RFF action-level dataset. For instance, a small number of deleted sites have ongoing actions in early stages of the pipeline. We close out all of these underway actions (4 RI/FSs, 3 RAs, and 2 LTRAs), since by definition, deleted sites have attained remedial goals and should not require future cleanup actions, Fund-lead or otherwise. There also are ongoing actions at sites designated by EPA as construction-complete; we assume that these underway actions will not be followed by subsequent cleanup actions. We also close out 12 RI/FSs and 8 RDs that, according to CERCLIS data, were under way at the end of FY 1999 at construction-complete sites listed as final. We do not close out any of the 141 RAs or 123 LTRAs under way at construction-complete NPL sites, since by definition, a site can be deemed construction-complete even if its last RA is still under way or operable units are undergoing LTRA.

There are 26 operable units at nonmega sites that had actions under way at the end of FY 1999 for which we do not have planned completion dates because the information is not included in the planning data provided by EPA. Of these 26 operable units, 6 have future planned actions, so we use the planned start date of the future planned action as the planned completion date for the action under way. For the remaining 20 operable units, we use our own average action durations to assign planned completion dates. At 1 operable unit, to determine the planned completion date, we simply add the appropriate duration, in rounded fiscal quarters as shown in Table G-2, to the fiscal year and quarter that the action started. At the remaining 19 operable units, actions had already been under way longer than our calculated average duration. For these actions, we assign planned completion dates by adding half the average duration to the first quarter of FY 2000. Since we do not have sufficient information to determine how far along these actions are, we assume they are half completed and thus have half of their duration remaining. These half-durations are also shown in Table G-3 and are applied according to whether the action is under way at a teenager or nonteenager site, for reasons discussed in the next subsection.

Of the 16 operable units that had no actions completed or under way at the end of FY 1999, we project that 13 will go through all the remedial pipeline phases, each beginning at the RI/FS phase. We drop the remaining 3 operable units from our baseline dataset, since the sites are already construction-complete. We also add 70 new operable units not included in the RFF action-level dataset—60 at which EPA has planned a future action, and 10 to reflect EPA's expected number of operable units at each site.

Thus, at the 1,133 nonmega sites on the NPL at the end of FY 1999, there were a total of 1,794 operable units. After we account for operable units at which remedial pipeline activity has been completed or the RODs call for no further action, there were 1,019 operable units with one or more ongoing or future actions yet to be completed at the end of FY 1999. Of these operable units, 114 have only LTRAs, meaning that most of the remedial pipeline costs have already been incurred. Of the 905 operable units at earlier stages of the pipeline, 141 are at construction-complete sites, which suggests that most of their remedial pipeline costs have already been incurred. The next step in the model is to project these 1,019 operable units through the rest of their respective pipeline phases, as discussed below.

Table G-3. Durations and Half-Durations Used in Model for Remedial Pipeline Phases at Nonmega Sites on Current NPL

	Teenager site				Nonteenager site			
	Duration		Half-duration		Duration		Half-duration	
	Years	Quarters	Years	Quarters	Years	Quarters	Years	Quarters
RI/FS	5.00	20	2.50	10	3.75	15	1.75	7
Remedial design	2.75	11	1.25	5	2.00	8	1.00	4
Remedial action	3.75	15	2.00	8	3.00	12	1.50	6

Note: Half-durations were used for actions under way at the end of FY 1999 that did not have planned completion dates.

Pipeline Projections

There are three components to projecting operable units through the pipeline: (1) estimating the number of future actions, (2) estimating whether the actions are likely to be Fund-lead or PRP-lead, and (3) forecasting when these actions are likely to occur.

In estimating the number of future actions, our basic assumption is that each operable unit will move through the remaining phases of the remedial pipeline. However, at some operable units, EPA determines that no further action is necessary. This determination usually is made when the record of decision is signed. Moreover, some operable units are divided into two or more operable units for cleanup purposes, typically between the RI/FS stage and the RD stage or between the RD stage and the RA stage, which we refer to as OU growth.

We assume that 14.85% of all future RI/FSs will result in NFA RODs—the average number of NFA RODs at nonmega sites from FY 1997 through FY 1999. Further, we assume that 5.49% of operable units will subdivide into two operable units between the RI/FS and RD stage and 1.6% of operable units will subdivide into two operable units between the RD and RA stage. These percentages are based on historical OU growth, which we determine by calculating the percentage change in the number of operable units from the RI/FS to RD stage and the RD to RA stage at sites that were construction-complete at the end of FY 1999. Since it is not possible to predict which operable units will have an NFA ROD or OU growth, we account for their incidence by adjusting our unit costs, as explained later in this appendix.

Another action we have to account for in the model is the future incidence of LTRAs. Again, we are not able to identify specific operable units that will require LTRAs, in large part because LTRAs are not necessarily included in EPA planning data. As discussed in Chapter 4, we assume that 45% of operable units have or will have a ROD that calls for groundwater or surface water restoration. Thus, to account for future LTRA costs, we assume that LTRAs will be conducted after the RA is completed at 45% of the operable units at which there is or will be a Fund-lead RA. Similarly, for 45% of the operable units at which there is or will be a PRP-lead RA, we assume that PRP LR (i.e., LTRA paid for by PRPs) will be conducted after the RA is completed, although the only Trust Fund dollars associated with these actions will be for oversight. We take into account operable units at which LTRAs and PRP LRs are already scheduled, according to EPA planning data.

Although we assume that no future RI/FSs, RDs, or RAs will be conducted at construction-complete sites, we do assign LTRAs to some operable units at construction-complete sites with RA actions under way at the end of FY 1999. We do not assign LTRAs to any of the 16 projected long RAs (i.e., future RAs with a planned duration of at least 10 years).

EPA designated all ongoing actions at the end of FY 1999 in the baseline dataset (932 actions) and all planned actions from Agency planning data for the next action at each operable unit (547 actions) as either Fund-lead or PRP-lead.* This information is included in the CERCLIS data that we use to create the RFF action-level dataset. Because most operable units at nonmega sites on the current NPL have progressed at least partway through the remedial pipeline, we already have lead information for a large number of the future actions to be conducted at sites on the NPL at the end of FY 1999. Where we do not have lead data from EPA, we have to make assumptions about whether these actions will be Fund-lead or PRP-lead. In total, we assign a lead to 829 future actions for operable units at sites on the current NPL. In most cases, the lead we assign is the same as the preceding action. For instance, 287 of the 829 future actions are LTRAs, and each LTRA is assigned the same lead as the preceding RA.

We assign leads for future actions where we do not have lead information from EPA by using historical event lead data from FY 1991 through FY 1999 provided to RFF by the EPA Office of Enforcement and Compliance Assurance (OECA). This period was chosen to reflect the Superfund program under the enforcement-first policy. We assign leads so that the overall percentage of future Fund-lead actions at all stages of the pipeline matches the historical percentages, shown in Table G-4. We assign these leads for each of the 11 RFF site types.† For example, of the 127 future RD actions for which we designate leads, 18 occur at chemical manufacturing sites. Table G-4 indicates that since FY 1991, 32% of RDs at chemical manufacturing sites have been Fund-lead; therefore we designate 6 of these RDs as Fund-lead and the remaining 12 as PRP-lead. Future LTRAs are assigned the same lead as the preceding RAs, because rarely has there been a change in lead between these two phases of the remedial pipeline. Based on these assumptions regarding the leads of future actions and our

---

*As discussed in Appendix F, we reorganized EPA lead information for the purposes of our analyses.

†At recycling sites, the rate of future PRP participation at the RD, RA, and LTRA stages of the remedial pipeline may not be consistent with historical percentages because of the liability exemptions set forth in the Superfund Recycling Equity Act of 1999. In the model there are a total of 7 RDs, 22 RAs, and 21 LTRAs at 28 sites that are designated PRP-lead and could be affected by this change.

Table G-4. Remedial Pipeline Action Lead History, by Type, at Nonmega Sites, FY 1996–FY 1999 (Percentage)

Type of site	RI/FS		Remedial design		Remedial action	
	Fund-lead	PRP-lead	Fund-lead	PRP-lead	Fund-lead	PRP-lead
Chemical manufacturing	45%	55%	32%	68%	35%	65%
Oil refining	67%	33%	25%	75%	21%	79%
Mining	40%	60%	83%	17%	67%	33%
Wood preserving	78%	22%	57%	43%	44%	56%
Coal gasification and other industrial	55%	45%	31%	69%	29%	71%
Recycling	58%	42%	18%	82%	19%	81%
Captive waste handling and disposal	38%	62%	14%	86%	11%	89%
Noncaptive waste handling and disposal	25%	75%	12%	88%	11%	89%
Transportation	88%	13%	11%	89%	29%	71%
Contaminated areas	73%	27%	36%	64%	37%	63%
Miscellaneous	86%	14%	49%	51%	41%	59%

Note: Percentages may not add to 100% because of rounding.
Source: Data provided to RFF by EPA, October 2000.

pipeline projections, Table G-5 shows the leads of actions under way and not yet started at nonmega sites, as well as the lead of actions that have been completed, according to the RFF action-level dataset.

We obtained from EPA the planned completion dates for most ongoing actions at NPL sites as well as the planned start dates for most future actions. These data are used to project the timing of both Fund-lead and PRP-lead actions, and ultimately to determine when to assign costs for Fund-lead actions. For each future cleanup action for which we do not have planned completion or planned start dates, we estimate the timing of these actions using average durations calculated for each phase of the remedial pipeline. First, we use as the start date the completion date of the preceding action, and we use as the completion date the start date plus the average duration for each type of action. Each action is assigned one of two durations, depending on whether it is conducted at a teen-

Table G-5. Summary of Remedial Pipeline Actions, by Type, at Nonmega Sites on Current NPL, End of FY 1999

	RI/FS		Remedial design		Remedial action		LTRA	
	Fund-lead	PRP-lead	Fund-lead	PRP-lead	Fund-lead	PRP-lead	Fund-lead	PRP-lead
Actions completed	610	605	321	690	208	492	3	10
Actions under way	108	107	63	126	118	265	48	97
Actions not yet started	36	37	137	236	155	399	123	253
Total	**754**	**749**	**521**	**1,052**	**481**	**1,156**	**174**	**360**

Note: PRP-lead LTRA is also referred to as PRP LR.

ager or nonteenager site. As explained in Appendix F, we round durations to the nearest quarter. Table G-6 shows the durations we use in the model.*

There are two exceptions to this rule. The first regards the time from the date a site is proposed to the NPL to the start of the first RI/FS, which depends on whether the RI/FS is Fund-lead or PRP-lead. We find in our analyses of CERCLIS data that if the first RI/FS at a nonmega site is Fund-lead, on average it starts about six months after the site is proposed to the NPL, versus an average 1.21 years if the first RI/FS is PRP-lead. This distinction turns out to be almost insignificant for our model for current NPL sites. Because cleanup work has already begun at most operable units at sites on the current NPL, we found only 21 future RI/FSs at operable units where no remedial pipeline action had started by the end of FY 1999. Moreover, 19 of these 21 operable units are at sites proposed to the NPL too early to use this duration, since using our durations would give them planned start dates before FY 2000. Instead, these 19 RI/FSs are assigned planned start dates in the first quarter of FY 2000. This most likely accelerates remedial pipeline activity at these operable units relative to EPA's actual plans; however, it is consistent with our underlying assumption that there are no resource constraints in our estimates of future costs.

The second exception concerns LTRAs, which represent the first 10 years of operation and maintenance of groundwater or surface water restoration remedies at operable units with a Fund-lead RA. As mandated by the National Contingency

---

*EPA planning data tend to be optimistic, and since we use these data to project the timing of future actions—both directly and in developing our average durations—our cost estimates may be slightly front-loaded.

Table G-6. Durations Used in Model for Remedial Pipeline Phases at Nonmega Sites on Current NPL

Pipeline phase	Teenager site		Nonteenager site	
	Years	Quarters	Years	Quarters
NPL proposal date to first RI/FS (Fund-lead)[a]	—	—	0.50	2
NPL proposal date to first RI/FS (PRP-lead)[a]	—	—	1.25	5
RI/FS	5.00	20	3.75	15
RI/FS completion to remedial design start	1.00	4	0.75	3
Remedial design	2.75	11	2.00	8
Remedial action	3.75	15	3.00	12
Long RA	20.50	82	20.50	82
LTRA	9.00	36	9.00	36
PRP LR	30.00	120	30.00	120

[a]Durations for this phase were not calculated for teenager sites for reasons detailed in Appendix F.

Plan (NCP), an LTRA is considered part of the RA for funding purposes, meaning that EPA pays 90% of the costs and the state pays the remaining 10%. (After this 10-year period, the state is responsible for 100% of the costs.) Notwithstanding the 10-year limit, we found the average duration of LTRAs to be about 9 years. At operable units where responsible parties are conducting or will conduct similar activities—coded PRP LR—we use an estimated duration of 30 years.

Lastly, as detailed in Appendix F, we use two sets of average durations for RA actions: one for RAs with durations of less than 10 years and another set for the small number of RAs with durations of 10 years and longer. As explained in Appendix F, the 10-year cutoff point is arbitrary. We incorporate the incidence of these so-called long RAs by assigning 4% (based on the percentage of RAs started after FY 1992 that had planned durations of 10 years or longer) of all future RA actions a duration of 20.5 years, or 82 quarters. The remaining 96% of future RA actions were assigned durations of 15 quarters or 12 quarters, depending on whether they were to be conducted at teenager or nonteenager sites.

Pipeline Costs
After we project each operable unit at each nonmega site through the remedial pipeline, we estimate total future costs by applying our unit costs to each indi-

vidual Fund-lead action and summing across all operable units by fiscal year. Actions with future costs include all Fund-lead RI/FSs, RDs, RAs, and LTRAs that either were under way at the end of FY 1999 or have planned start dates in FY 2000 or beyond. PRP-lead actions are not assigned direct cleanup costs.

We assign unit costs for nonmega sites by the type of remedial pipeline action (RI/FS, RD, RA, LTRA) and—for extramural costs—by RFF site type (LTRAs are not varied by site type). As discussed in Appendix F, we develop our unit costs for RI/FSs, RDs, and RAs by analyzing historical expenditure data from IFMS. For extramural unit costs, we use data from FY 1992 through FY 1999 according to RFF site type; for intramural unit costs, we use data from FY 1996 through FY 1999 *not* varied by site type. The unit costs that we use are shown in Table G-7. LTRAs at nonmega sites are given an average annual cost of $400,000, which includes extramural costs only (intramural costs associated with LTRAs are captured as part of site-specific response support, discussed later in this appendix).

We do not assign the full unit cost for Fund-lead RI/FSs, RDs, and RAs that were under way at the end of FY 1999, since these actions were already partially completed. For these underway actions, we multiply the appropriate unit cost by the percentage of the action remaining, defined as the number of quarters between the start of FY 2000 and the planned completion date divided by the number of quarters between the start date and the planned completion date. We then distribute these adjusted unit costs uniformly over all future fiscal quarters between the first quarter of FY 2000 and the planned completion date.

For example, a Fund-lead RA at a nonmega oil refining site that started in the third quarter of FY 1998 and is projected to be completed in the second quarter of FY 2001 has 6 quarters remaining of a total duration of 12 quarters and therefore has 50% of its costs remaining. Thus, according to Table G-7, this RA has $2,255,456 of extramural costs and $21,350 of intramural costs left to be incurred, or $375,909 in extramural costs and $3,558 in intramural costs for each of the 6 remaining fiscal quarters from the first quarter of 2000 through the second quarter of 2001. Fund-lead LTRAs under way at nonmega sites at the end of FY 1999 are assigned costs of $100,000 for each fiscal quarter from the first quarter of FY 2000 through their planned completion dates.

As noted previously, we adjust our cost estimates for those future Fund-lead RI/FSs, RDs, RAs, and LTRAs at nonmega sites that had not yet started by the end of FY 1999 and that are *not* included in EPA's planning data to take into account the incidence of NFA RODs and OU growth. Table G-8 shows these unit cost adjustments. We do not adjust future RI/FS unit costs because the impact of NFA RODs and OU growth occurs only after this point in the

Table G-7. Unit Costs of Fund-Lead Remedial Pipeline Actions, by Type, at Nonmega Sites on Current NPL (1999$)

Type of site	RI/FS	Remedial design	Remedial action
Extramural unit costs			
Chemical manufacturing	711,383	826,128	5,199,644
Oil refining	552,938	905,203	4,510,912
Mining	2,222,175	1,172,790	10,325,396
Wood preserving	752,248	1,005,443	10,553,716
Coal gasification and other industrial	1,110,903	814,660	5,655,922
Recycling	1,163,172	1,079,953	4,418,436
Captive waste handling and disposal	439,805	776,741	3,667,780
Noncaptive waste handling and disposal	1,039,977	1,197,684	5,940,744
Transportation	559,387	447,248	10,452,720
Contaminated areas	970,140	648,690	4,058,127
Miscellaneous	1,171,268	608,855	3,198,050
Intramural unit costs			
All nonmega sites	56,757	27,392	42,699

cleanup process. We multiply the unit costs for the future RDs we project by 89.8%, a factor that incorporates both NFA RODs and OU growth between the RI/FS and RD phases of the pipeline. Future RAs are multiplied by one of two factors. If the actions are to follow future RDs that we project, we account for NFA RODs and both types of OU growth by multiplying unit costs by 91.3%. However, if the RAs are to follow ongoing, completed, or future RDs planned by EPA, NFA ROD incidence and the first OU growth are not applicable, so we multiply unit costs by 101.6% to take into account OU growth between the RD and RA phases of the pipeline. The annual costs of the future LTRAs that we project are multiplied by the same factor as the RAs that preceded them.

After we adjust unit costs to account for NFA RODs and OU growth, we distribute the unit cost dollars evenly, by quarter. For future Fund-lead RI/FSs, RDs, and RAs, we distribute the adjusted extramural and intramural unit costs uniformly across all quarters during which those actions are projected to be under way. The exceptions to this rule are the long RAs, which have average

Table G-8. Unit Cost Adjustments of Remedial Pipeline Actions at Current NPL Sites to Incorporate NFA RODs and OU Growth

Future action projected by RFF	Adjustment factor (%)
RI/FS	100.0%
Remedial design	89.8%
Remedial action (including long RA)	
that follows a future remedial design projected by RFF	91.3%
that does not follow a future remedial design projected by RFF	101.6%
LTRA	
that follows a future remedial action projected by RFF	
and that remedial action follows a future remedial design projected by RFF	91.3%
and that remedial action does not follow a future remedial design projected by RFF	101.6%
that does not follow a future remedial action projected by RFF	100.0%

durations of 20.5 years. Since most of the costs at Fund-lead long RAs are front-loaded, because these costs relate primarily to the construction of remedies, we allocate 75% of the unit costs uniformly over the first 3 years of the action and allocate the remaining 25% uniformly over the remaining 17.5 years. The average annual cost for LTRAs is divided into average quarterly costs, which are assigned to each quarter between the start and completion date of the action.

Total Cost of Fund-Lead Actions at Current NPL Sites

After extramural and intramural unit costs are assigned to Fund-lead actions at operable units, we sum these dollars across all sites, by fiscal year, to obtain our estimates for future direct cleanup costs for Fund-lead actions at non-mega sites on the NPL at the end of FY 1999. We then add these estimates to the future direct cleanup costs for mega sites on the NPL at the end of FY 1999 to obtain our total estimates for the future direct costs of Fund-lead cleanup actions at current NPL sites.

These results represent the base case for Fund-lead actions for the current NPL. As discussed in Chapter 7, we also estimate the future cost of Fund-lead cleanups at current NPL sites under two other scenarios—a low case and a high case—by making adjustments to some of the assumptions and inputs described above.

For the low case, we incorporate a 20% reduction in the cost of future RAs (extramural costs only) to reflect recent projections by EPA that the cost of conducting RAs are declining. We apply this cost savings to all future RAs at mega sites by taking 20% off the estimates provided by the RPMs in the RFF mega site survey. For nonmega sites, we discount RA unit costs by 20%. All other components of the model are left unchanged.

For the high case, we use RFF unit costs in lieu of the RPMs' estimates for Fund-lead actions at mega sites to account for the possibility that the RPMs have underestimated the true cost of future Fund-lead actions at mega sites. We discovered that, in aggregate, the RPMs' estimates are lower than if we use the average unit costs we calculated for mega sites based on recent historical expenditures. We also included in our high case the cost of additional Fund-lead actions that RPMs said might be necessary at their sites but for which they did not have cost estimates. For these actions, we distribute unit costs (extramural and intramural) evenly from FY 2005 through FY 2009. (These include four RI/FSs, five RDs, six RAs, and three LTRAs.) We do not assign the cost of these additional actions until FY 2005 on the theory that the actions are unlikely to be conducted in the next few years if the RPMs had so little information about them. All other components of the model in the high-case scenario are the same as in the base case.

Construction-Complete Projections

At the end of FY 1999, 597 (48%) of the 1,245 final and deleted sites on the NPL had *not* been designated construction-complete. To both track the progress of cleanup at NPL sites and estimate the timing of five-year reviews required by EPA policy (see discussion in the next section), it is necessary to estimate the fiscal year in which these 597 sites will reach construction completion. The CERCLIS data we use to create the RFF site-level dataset include an estimated construction-complete date (by fiscal year) for such sites, but we decided to estimate construction-complete dates that would be consistent with our model. Our estimates are based, in part, on EPA responses to the RFF mega site survey (for mega sites), EPA planning data, and the average durations we calculated for remedial pipeline actions (for nonmega sites), since these are the inputs that determine site progress in our model.

For mega sites, as part of the RFF mega site survey, remedial project managers were asked to indicate what cleanup actions would be conducted at their sites.

From this information we identify the last RA to be conducted at each mega site and estimate that the construction-complete fiscal year will be one year less than the date of expected completion of the last RA. The reason we select the year before the RA is expected to be complete is that RAs generally continue for at least one year after they are fully implemented while EPA ensures that the remedy is operating properly. This time is referred to as the operational and functional period. In a few instances our approach suggests that a non-construction-complete site should already be construction-complete, in which case we assume that this site will reach construction completion in FY 2000.

We assume that any nonmega site whose last RA was completed or that was deleted from the NPL at the end of FY 1999 is construction-complete in FY 2000. If the last RA has yet to occur, we assume that the site will reach construction completion one year before this last RA is complete, unless the last RA is a long RA, in which case we assume that the site will reach construction completion three years after the start date of this last RA. For nonmega sites whose last RA was already under way at the end of FY 1999, we assume that the site will become construction-complete one year before the fiscal year in which this last RA is expected to be finished or the actual start date (in fiscal year) of this last RA plus 10 years, whichever is earlier. (This takes into account last RAs under way that already have long durations.) As for mega sites, in a few instances our approach suggests that a non-construction-complete site should already be construction-complete, in which case we assume that this site is construction-complete in FY 2000.

Lastly, nine sites—three mega sites and six nonmega sites—do not have past or future RAs. For six of these sites, we use EPA's estimated construction-complete fiscal year. For the other three, we do not have an estimated construction-complete fiscal year from EPA and assume they will reach construction completion in FY 2000.

Five-Year Review Projections

Five-year reviews are required at any site where the selected RA(s) leaves hazardous substances, pollutants, or contaminants remaining above levels that allow for unlimited use and unrestricted exposure. Using an estimate of $25,000 for nonmega sites and $50,000 for mega sites (see Chapter 4 and Appendix F), we estimate the total cost EPA will incur for conducting five-year reviews from FY 2000 through FY 2009. Our model includes cost projections only for five-year reviews to be performed at sites on the NPL at the end of FY 1999. We assume that sites added to the NPL in FY 2000 and beyond will not require five-year reviews from FY 2000 through FY 2009, based on our average durations for the timing of actions at future NPL sites. It should be noted that although five-year reviews may be required at non-NPL sites (i.e., sites proposed to the NPL, removal-only sites, and NPL equivalent sites) if hazardous substances remain above levels that allow for unlimited use and unrestricted exposure, the cost of these reviews is not included in our estimates.

EPA provided RFF with a number of datasets containing information about five-year reviews (summarized in Appendix C), which we use to estimate when the Agency will incur costs for performing reviews at the 1,245 NPL sites listed as final or deleted at the end of FY 1999. Rather than select an exact due date, we determine in which fiscal year the next five-year review will be due and assume that subsequent reviews will occur at five-year intervals from this year. We assume that each five-year review will be completed in the same fiscal year it is started. Moreover, we assume that EPA will incur the cost in the same fiscal year in which the five-year review is performed.

For 757 of the 1,245 sites on the NPL at the end of FY 1999, EPA provided RFF with sufficient information to determine when the next five-year review will be conducted. EPA designated 129 of these sites as walk-aways—sites that will not require any future five-year reviews. We do not assign any five-year review costs to these 129 sites. At another 112 sites, a five-year review was overdue at the end of FY 1999. Since EPA has committed to eliminating this backlog by the end of FY 2002 (and the Agency's progress toward this goal in FY 2000 suggests that it will meet this commitment), we assume that a third of these reviews will be completed each year from FY 2000 through FY 2002, and we assume that subsequent reviews will be performed every five years thereafter. If the original due date for an overdue five-year review was in FY 1998 or FY 1999, we assume that the overdue five-year review will be conducted during the FY 2000 to FY 2002 period, but that its subsequent five-year review will be conducted five years from the original due date. This is consistent with EPA guidance.

At 288 sites, EPA data indicated that a five-year review was to be completed in FY 2000, FY 2001, or FY 2002. We assign the costs for these sites in the fiscal year in which the reviews are to be completed and assume that subsequent reviews at these sites will be performed every five years thereafter. At the final group of 228 sites for which we have data from EPA, at least one five-year review had already been conducted as of the end of FY 1999. According to EPA data, the Agency did not plan to perform a subsequent review in FY 2000, FY 2001, or FY 2002 for any of these sites. Thus, for these sites, we assume that the next review will occur either five years from the original due date or five years from the completion date of the last review, whichever is earlier.

For the other 488 sites on the current NPL, the data EPA provided do not contain sufficient information for us to determine the timing of future five-year reviews. At these sites, we assign due dates by fiscal year according to the rules EPA set forth in the NCP and in Agency guidance, which indicate that the due dates for five-year reviews are contingent on the type of review—statutory or policy. Section 300.430(f)(4)(ii) of the NCP specifies that statutory reviews should be conducted "no less often than every five years after initiation of the selected remedial action." Accordingly, we assume that statutory reviews are due five years after the first RA starts at a site and every five years thereafter. Policy reviews are to be conducted five years after a site is designated by the Agency as construction-complete and every five years thereafter. (See the previous section of this appendix for an explanation of the methodology we use to determine when a site will be construction-complete.)

The data provided by EPA include information on the type of review—statutory or policy—for 879 of the 1,245 final and deleted sites on the NPL at the end of FY 1999. We assume that the rest of the sites, excluding those EPA designated as walk-away sites, will require five-year reviews by statute. Thus, for the 488 sites for which we must determine when the next five-year review will occur, we use the statutory or policy rules described above.

The only exceptions are for two small sets of sites. First, 14 sites had five-year reviews due before FY 2000, but it was not clear from EPA data whether the reviews had been completed or were overdue. For these sites, we assume the next five-year reviews will be conducted five years from their original due dates. Second, there are seven sites for which we cannot determine the appropriate timing for the five-year reviews. Three of these sites were deleted from the NPL but have not been designated construction-complete and never had RAs. We assume that five-year reviews will be conducted five years from the date these sites were deleted from the NPL. At the other four sites, the last action recorded was either a removal or a discovery in FY 1982, and the site never had RAs and has not been designated construction-complete. We assume these sites are walk-aways and will not require five-year reviews.

Cost of Fund-Lead Actions at Future NPL Sites

We estimate the cost of Fund-lead actions under three future NPL scenarios. In this section of the appendix, we first discuss our overall approach and review the main assumptions underlying each scenario. Second, we briefly describe the three building blocks used in our estimates for all the scenarios. Lastly, we explain the methodology we use to develop our cost estimates.

Overall Approach and Three Future Scenarios

The approach we take to estimate the cost of Fund-lead actions at future NPL sites differs from that for the current NPL. Although we use some of the same building blocks regarding durations and unit costs and make our estimates at the operable unit level, we in effect create "generic" future NPL sites. The main reason for estimating costs in this manner is that we know less about future NPL sites than about the current sites. For instance, it is not possible to predict the annual distribution of site types or the number of operable units at each site. Thus, we do not assign site types, and we assume that each site has the same number of operable units—3.8 operable units at a mega site and 1.6 at a nonmega site.*

We base our estimates of the future cost of Fund-lead actions at sites added to the NPL from FY 2000 through FY 2009 on three key variables: the total number of sites added to the NPL each year, the number of these sites that will be mega sites, and the percentage of actions at these sites (mega sites and nonmega sites) that will be Fund-lead. As noted in Chapter 5, there is considerable uncertainty regarding these variables. To address this uncertainty, we estimate the cost under three scenarios, varying the number of total listings per year, the proportion of mega versus nonmega sites, and the proportion of remedial pipeline actions that are Fund-lead versus PRP-lead. To reflect actual listings in FY 2000, we assume in each scenario that 36 sites are added to the final NPL in FY 2000, two of which are mega sites.† Table G-9 summarizes the three scenarios.

Building Blocks

Before discussing the methodology we use to estimate the future cost of Fund-lead actions at future NPL sites, it is useful to discuss the building blocks that are common to all the scenarios. The first regards the duration of remedial pipeline actions. We assume that the average duration of a future

---

*The average number of operable units for mega sites and nonmega sites is based on the average for these sites calculated from all final and deleted nonfederal sites on the NPL at the end of FY 1999.

†Technically, 37 sites were added to the NPL in FY 2000, but one, Georgia-Pacific Corp. in North Carolina (Region 4), was withdrawn from the NPL later that fiscal year.

Table G-9. Summary of RFF Future NPL Scenarios, FY 2001–FY 2009

	Base case	Low case	High case
Number of sites listed each year			
Total	35	23	49
Mega	2	1	3
Nonmega	33	22	46
Percentage of actions that are Fund-lead at mega sites			
RI/FS	80%	80%	80%
Remedial design	35%	35%	35%
Remedial action	25%	25%	35%
Percentage of actions that are Fund-lead at nonmega sites			
RI/FS	75%	75%	75%
Remedial design	40%	30%	40%
Remedial action	40%	30%	40%

action will depend on whether the action is conducted at a mega or nonmega site. As shown in Table G-10, the duration of each action is rounded to the nearest quarter. (Appendix F discusses the calculation of these durations and the conversion to rounded quarters.) As with the current NPL, we do not distinguish between Fund-lead and PRP-lead actions in durations, except for the small percentage of remedial actions that include groundwater or surface water restoration (LTRAs and PRP LRs).

The second building block is the cost of Fund-lead remedial pipeline actions. For future NPL sites we cannot use site-specific information or predict what types of sites will be added to the NPL each year. (We did try, unsuccessfully, to gather this information in our interviews with state and EPA officials.) Thus, in estimating the cost of future NPL sites, we vary unit costs only by whether the action is to be conducted at a mega or nonmega site. Table G-11 shows the extramural and intramural unit costs we use in the model for each remedial pipeline phase. These unit costs are used in all the scenarios, with one exception. We assume that in the low case for future NPL sites (as for current sites), the extramural costs of RAs will be 20% less than our unit costs. We do not assign any direct cleanup costs between the time of the NPL final listing date and the start of the first RI/FS, or between the RI/FS and the RD. We also

Table G-10. Durations Used in Model for Remedial Pipeline Phases at Future NPL Sites

Pipeline phase	Mega sites		Nonmega sites	
	Years	Quarters	Years	Quarters
NPL final date to RI/FS start	0.25	1	0.00	0
RI/FS	5.00	20	3.50	14
RI/FS completion to remedial design start	0.25	1	0.75	3
Remedial design	2.00	8	2.00	8
Remedial action	3.25	13	3.00	12
Long RA	20.50	82	20.50	82
LTRA	10.00	40	9.00	36
PRP LR	30.00	120	30.00	120

Table G-11. Unit Costs of Fund-Lead Remedial Pipeline Actions at Future NPL Sites (1999$)

Pipeline phase	Mega sites		Nonmega sites	
	Extramural	Intramural	Extramural	Intramural
RI/FS	2,479,220	82,464	991,011	56,757
Remedial design	3,886,939	49,808	884,330	27,392
Remedial action	30,364,636	120,278	5,756,692	42,699
LTRA (per year)	2,000,000	—	400,000	—

assume (as for the current NPL) that long RAs will have the same unit cost as other RAs, and that LTRAs are given costs on an annual basis.

By using the date a site is listed as final as our starting point for estimating costs, we assume that no money will be spent on Fund-lead actions from the date a site is proposed to the NPL until it is listed as final. Even though RI/FSs often begin when a site is proposed to the NPL, we make this assumption in order to estimate the costs only of sites that become final (i.e., we exclude costs incurred at proposed sites that will not be made final). Because of the way we calculate extramural and intramural unit costs, the money spent on Fund-lead actions at *proposed* NPL sites is taken into account but not allocated until after the site is listed as final. Thus, the timing of these costs is slightly compressed.

As for the current NPL, we adjust the unit costs of remedial pipeline actions at future sites to allow for the incidence of NFA RODs and OU growth. Since we do not use site-specific information, we also need to adjust the unit costs for actions conducted at mega sites. Though we use the same rate of incidence for OU growth, as Table G-12 shows, we calculate a new rate of incidence for NFA RODs at mega sites to take into account the difference between the frequency of NFA RODs at these sites compared with nonmega sites. We assume in our model that 8% of all RI/FSs conducted at mega sites will result in NFA RODs— the average incidence of NFA RODs at mega sites from FY 1997 through FY 1999. For nonmega sites, we assume that 14.85% of all RI/FSs conducted will result in NFA RODs—the average incidence of NFA RODs at nonmega sites from FY 1997 through FY 1999.

Using these rates of incidence, we adjust the unit costs of remedial pipeline actions at future sites up or down so that the total cost reflects the changed number of actions, even though the number of actual operable units does not change. The unit cost adjustment factors attributable to NFA RODs and OU growth are shown in Table G-13. For a more detailed explanation of this method of incorporating changes to the number of operable units in different stages and making adjustments to unit costs, see the earlier section of this appendix that discusses the current NPL.

The last building block is the proportion of remedial pipeline actions at future NPL sites that will be Fund-lead. As discussed above in reference to the current NPL, operable units often change lead as they progress through the pipeline, so it is incorrect to assume that all actions taken at an operable unit will be Fund-lead or PRP-lead. In general, there have historically been more Fund-lead RI/FSs than PRP-lead RI/FSs; the opposite is true for RDs and RAs. Table G-14 shows the breakdown of recent historical (FY 1996 through FY 1999) action-lead percentages by pipeline stage, for both mega and nonmega sites.

Table G-12. Changes in the Number of Operable Units for Future NPL Sites (Percentage)

	Rate of incidence	
	Mega sites	Nonmega sites
NFA ROD	8.00%	14.85%
OU growth, RI/FS to remedial design	5.49%	5.49%
OU growth, remedial design to remedial action	1.60%	1.60%

Table G-13. Unit Cost Adjustments for Remedial Pipeline Actions at Future NPL Sites to Incorporate NFA RODs and OU Growth (Percentage)

Type of action	Mega sites	Nonmega sites
RI/FS	100.00%	100.00%
Remedial design	97.05%	89.82%
Remedial action (including long RA)	98.60%	91.26%
LTRA	98.60%	91.26%

Table G-14. Remedial Pipeline Action Lead History, FY 1996–FY 1999 (Percentage)

Type of action	Mega sites		Nonmega sites	
	Fund-lead	PRP-lead	Fund-lead	PRP-lead
RI/FS	80%	20%	75%	25%
Remedial design	35%	65%	30%	70%
Remedial action	25%	75%	30%	70%

Note: Percentages have been rounded.

Methodology for Estimating the Cost of Future NPL Sites

We assume that in each year from FY 2000 through FY 2009, a certain number of sites will be added to the NPL (as final), a small portion of which will be mega sites. (As noted above, in each scenario we assume that 36 sites, including 2 mega sites, are added to the NPL in FY 2000, which reflects the actual listings in that year.) We then make a number of assumptions to estimate the cost EPA is likely to incur for conducting Fund-lead actions at these sites from FY 2000 through FY 2009.

We estimate the cost of Fund-lead actions at future NPL sites at the site level. In essence, we estimate the cost of cleaning up a generic mega site and a generic nonmega site and then multiply the annual costs associated with these sites by the number of mega sites and nonmega sites listed each year. We estimate the cost of these generic sites based on the building blocks described above, adjusted when necessary to reflect the different assumptions in each scenario.

To estimate the cost of Fund-lead actions at a generic mega site and a generic nonmega site, we first consider the cost associated with a typical operable unit at each type of site. For each operable unit, we use the durations and unit costs discussed above and shown in Table G-10 and Table G-11, respectively. We

Table G-15. Possible Operable Unit Paths at Future NPL Sites

Path	Likelihood	Remedial pipeline
RA without LTRA	51%	NPL final date to RI/FS → RI/FS → RI/FS-to-RD→ RD → RA
RA with LTRA	45%	NPL final date to RI/FS → RI/FS → RI/FS-to-RD→ RD → RA →LTRA
Long RA	4%	NPL final date to RI/FS → RI/FS → RI/FS-to-RD→ RD → Long RA

assume that only Fund-lead actions will have direct cleanup costs,[*] so we multiply the unit cost for each phase of the remedial pipeline by the appropriate Fund-lead percentages, which vary for mega versus nonmega sites and under each scenario. Table G-9 shows the percentages for the three scenarios. Moreover, to account for NFA RODs and OU growth, we multiply the unit costs for each phase of the remedial pipeline by the appropriate adjustment factor shown in Table G-13. We then apportion these unit costs uniformly over the pipeline phase duration.

We also assume that LTRAs or PRP LRs will follow only 45% of RAs and that 4% of RAs will be long RAs. Thus, there are three potential paths for an operable unit: RA followed by LTRA, RA not followed by LTRA, and long RA. The likelihood of each of these three paths and their associated remedial pipelines are shown in Table G-15. We first estimate the future annual costs for these three possible paths and then calculate the average cost path for a single operable unit by averaging the costs of the three paths together, weighting each by its respective likelihood.

At this point, we have created an average operable unit for both a mega site and a nonmega site. The next step is to move from the operable unit level to the site level to create a representative site. To do this, it is necessary to account for both the number of operable units at each kind of site and OU lag (the time between the start of remedial pipeline activity at an operable unit and the start of remedial pipeline activity at each successive operable unit at the site). We assume that there are 1.6 operable units at a nonmega site (the average number of operable units at all nonmega sites listed as final or deleted on the NPL at the end of FY 1999). To account for OU lag, we assume that the RI/FS will not

---

[*]Costs associated with the oversight of PRP-lead cleanup actions are captured in another part of the model.

begin at a second operable unit until one year after the first.* Because we assume that each nonmega site has 1.6 operable units, the model estimates the costs for two average nonmega operable units (with start dates one year apart), multiplying the cost of each by 0.8 (50% of 1.6). We then sum the costs of these operable units to estimate the future annual cost of Fund-lead actions at one nonmega site.

We use the same approach for mega sites, which have an average of 3.8 operable units per site (the average number of operable units at mega sites listed as final or deleted on the NPL at the end of FY 1999). Because we assume that each mega site has 3.8 operable units, the model estimates the costs for three average operable units (with start dates one year apart), multiplying the cost of each by 1.27 (33% of 3.8). We then sum the costs of these operable units to estimate the future annual cost of Fund-lead actions at one mega site.

The final step in our approach is to multiply the cost estimates for the average mega site and average nonmega site by the number of mega sites and nonmega sites listed in a given year. This projection is calculated every year from FY 2000 through FY 2009, and the costs are then summed by the fiscal year in which they are incurred to obtain annual costs from FY 2000 through FY 2009.

Cost of the Rest of the Superfund Program

For reasons discussed in Chapter 7, we estimate the future cost of the non-Fund-lead remedial pipeline actions of the Superfund program (including five-year reviews) in one of two ways: adjusting costs based on FY 1999 expenditure levels by scaling factors or holding costs constant at FY 1999 expenditure levels. Each is discussed below.

Costs Adjusted by Scaling Factors

We use a number of scaling factors to adjust the annual costs associated with different components of the Superfund program. The scaling factors, in essence, are indicators of the workload or level of activity within the Superfund program. For each year we compare the estimated workload with the FY

---

*Based on final and deleted NPL sites proposed in FY 1992 or later, the average lag time between the first and the second RI/FS was 1.09 years. The average lag time between the second and the third RI/FS was 1.36 years.

1999 workload and adjust the associated costs for this part of the program up or down accordingly.

We use this approach because the cost of many components of the Superfund program are, at least in theory, connected to the level of activity of either a particular part of the program or to the program as a whole. Moreover, for many parts of the program, it is not possible to calculate a unit cost or average cost, as we did for Fund-lead actions at current and future NPL sites and five-year reviews. Because past Superfund expenditures have been driven primarily by budget decisions, it is not possible to empirically test how well our scaling factors truly measure different activities carried out within the program. Each scaling factor was developed, and the parts of the program we model using them were selected, after consultation with EPA and Department of Justice (DOJ), and in our judgement represent plausible indicators for estimating future costs of various Superfund program activities.

We discuss the scaling factors below, identifying how each is calculated and for which parts of the Superfund program it is used. Although the scaling factors are applied consistently for each scenario, the overall Superfund workload, of course, varies by scenario.

Adjusted PRP Actions

This scaling factor (Table G-16) is based on the total number of ongoing PRP-lead cleanup actions (RI/FSs, RDs, RAs, and LTRAs) at the end of each fiscal year.* To weight the level of work associated with actions at mega sites, each action at a mega site is multiplied by two. To weight the level of work associated with PRP LRs, we multiply each PRP LR by a factor of 0.2. This scaling factor is used to estimate the future cost of site-specific oversight activities.

Adjusted All-Actions

This scaling factor (Table G-17) is based on the total number of ongoing Fund-lead and PRP-lead actions (RI/FSs, RDs, RAs, and LTRAs) at the end of each fiscal year. To weight the level of work associated with different actions, we multiply each action at a mega site by two, and each LTRA and PRP LR by a factor of 0.2. This scaling factor is used to estimate the future cost of site-

---

*Since NFA RODs and OU growth are incorporated into unit costs, the number of actions in the model is slightly inflated (approximately 1% for the current NPL and 6% for the future NPL). Overall, we estimate that the error ranges from 2% to 4%, with small variations among the different scenarios. This error also applies to the adjusted all-actions scaling factor.

Table G-16. Scaling Factor for Adjusted PRP Actions: Three Scenarios, by Fiscal Year

Scenario	FY 1999	FY 2000	FY 2001	FY 2002	FY 2003	FY 2004	FY 2005	FY 2006	FY 2007	FY 2008	FY 2009
Adjusted PRP actions											
Low	865	752	721	710	674	603	537	493	446	424	402
Base	865	752	724	718	688	618	554	520	484	474	467
High	865	752	727	728	704	638	582	563	542	549	558
Scaling factor (%)											
Low	100%	87%	83%	82%	78%	70%	62%	57%	52%	49%	46%
Base	100%	87%	84%	83%	80%	71%	64%	60%	56%	55%	54%
High	100%	87%	84%	84%	81%	74%	67%	65%	63%	64%	65%

Table G-17. Scaling Factor for Adjusted All-Actions: Three Scenarios, by Fiscal Year

Scenario	FY 1999	FY 2000	FY 2001	FY 2002	FY 2003	FY 2004	FY 2005	FY 2006	FY 2007	FY 2008	FY 2009
Adjusted all-actions											
Low	1,357	1,235	1,171	1,170	1,082	968	869	806	725	688	648
Base	1,357	1,235	1,182	1,204	1,142	1,044	961	923	868	855	840
High	1,357	1,235	1,195	1,243	1,209	1,129	1,064	1,055	1,028	1,044	1,057
Scaling factor (%)											
Low	100%	91%	86%	86%	80%	71%	64%	59%	53%	51%	48%
Base	100%	91%	87%	89%	84%	77%	71%	68%	64%	63%	62%
High	100%	91%	88%	92%	89%	83%	78%	78%	76%	77%	78%

specific response support activities; site-specific community and state involvement activities; remedial program policy, management, and staff; response action management; and laboratory support.

NPL Activity

This scaling factor (Table G-18) is based on the total number of final and deleted nonfederal NPL sites at the end of FY 1999 (1,245), subtracting the total number of construction-complete sites at the end of the year, and adding the total number of new, nonfederal NPL sites added since the end of FY 1999. This provides an estimate of the total number of "active" sites—those that are not construction-complete. This scaling factor is used to estimate the future cost of the noncost recovery portion of DOJ enforcement and EPA enforcement activities, as well as the NPL share of enforcement policy, management, and staff.

DOJ Cost Recovery

This scaling factor (Table G-19) is based on the total number of last RA starts in the five-year period beginning seven years before the year for which the scaling factor will be used. This eight-year period accounts for the three-year lag before DOJ begins its cost recovery efforts at a site, and the five-year period it typically takes to pursue cost recovery. This scaling factor is used to estimate only the future cost of the cost recovery portion of DOJ enforcement.

EPA Cost Recovery

This scaling factor (Table G-20) is based on the total number of last RA starts in the five-year period beginning six years before the year for which the scaling factor will be used. This seven-year period accounts for the two-year lag before EPA begins its cost recovery efforts at a site, and the five-year period it typically takes to pursue cost recovery. This scaling factor is used to estimate only the future cost of the cost recovery portion of EPA enforcement.

"Total Dollars"

This scaling factor (Table G-21) is based on the sum of our annual estimates of the costs of the removal program, the remedial program, efforts at non-NPL sites, site assessment activities, other EPA offices and programs, and program staff, management, and support. It does not include our annual estimates for other agencies (National Institutes for Environmental Health Sciences and the U.S. Coast Guard) and the EPA Office of the Inspector General. This scaling factor is used to estimate the future cost of all program administration except Office of Administration and Resources Management (OARM) information systems and services.

Costs Held Constant

Certain components of the Superfund program are held at their FY 1999 expenditure levels from FY 2000 through FY 2009. We do this for costs that are fixed, such as the Office of Solid Waste and Emergency Response front office, and for others, such as the budget for the National Institute for Environmental Health Sciences, that are determined by Congress and are not directly related to the overall size of the Superfund program. Using the RFF IFMS dataset to determine FY 1999 expenditures, we hold constant the following components of the program at their FY 1999 expenditure level for each year from FY 2000 through FY 2009:

- removal program;
- site assessment activities;
- remedial program for non-NPL sites (e.g., NPL equivalent sites);
- Office of Solid Waste and Emergency Response front office;
- Office of Enforcement and Compliance Assurance front office and communications;
- National Enforcement Investigations Center;
- state programs and grants;
- Office of Research and Development;
- Office of the General Counsel;
- Technology Innovation Office;
- Chemical Emergency Preparedness and Prevention Office;
- U.S. Coast Guard;
- National Institute for Environmental Health Sciences;
- Office of the Inspector General; and
- OARM information systems and services.

Table G-18. Scaling Factor for NPL Activity: Three Scenarios, by Fiscal Year

Scenario	FY 1999	FY 2000	FY 2001	FY 2002	FY 2003	FY 2004	FY 2005	FY 2006	FY 2007	FY 2008	FY 2009
"Active" sites											
Low	597	547	550	529	491	453	402	336	313	307	256
Base	597	547	562	553	527	501	462	408	397	403	364
High	597	547	576	581	569	557	532	492	495	515	490
Scaling factor (%)											
Low	100%	92%	92%	89%	82%	76%	67%	56%	52%	51%	43%
Base	100%	92%	94%	93%	88%	84%	77%	68%	66%	68%	61%
High	100%	92%	96%	97%	95%	93%	89%	82%	83%	86%	82%

Table G-19. Scaling Factor for Department of Justice Cost Recovery: Three Scenarios, by Fiscal Year

Scenario	FY 1999	FY 2000	FY 2001	FY 2002	FY 2003	FY 2004	FY 2005	FY 2006	FY 2007	FY 2008	FY 2009
Lagged five-year running total of last remedial action starts (beginning seven years prior)											
Low	307	316	306	289	294	287	279	331	373	330	306
Base	307	316	306	289	294	287	279	331	373	330	306
High	307	316	306	289	294	287	279	331	373	330	306
Scaling factor (%)											
Low	100%	103%	100%	94%	96%	93%	91%	108%	121%	107%	100%
Base	100%	103%	100%	94%	96%	93%	91%	108%	121%	107%	100%
High	100%	103%	100%	94%	96%	93%	91%	108%	121%	107%	100%

Table G-20. Scaling Factor for EPA Cost Recovery: Three Scenarios, by Fiscal Year

Scenario	FY 1999	FY 2000	FY 2001	FY 2002	FY 2003	FY 2004	FY 2005	FY 2006	FY 2007	FY 2008	FY 2009
Lagged five-year running total of last remedial action starts (beginning six years prior)											
Low	316	306	289	294	287	279	331	373	330	306	318
Base	316	306	289	294	287	279	331	373	330	306	318
High	316	306	289	294	287	279	331	373	330	306	318
Scaling factor (%)											
Low	100%	97%	91%	93%	91%	88%	105%	118%	104%	97%	101%
Base	100%	97%	91%	93%	91%	88%	105%	118%	104%	97%	101%
High	100%	97%	91%	93%	91%	88%	105%	118%	104%	97%	101%

Table G-21. Scaling Factor for "Total Dollars": Three Scenarios, by Fiscal Year

Scenario	FY 1999	FY 2000	FY 2001	FY 2002	FY 2003	FY 2004	FY 2005	FY 2006	FY 2007	FY 2008	FY 2009
Total annual cost in billions of dollars (1999$)											
Low	1.34	1.27	1.26	1.37	1.37	1.24	1.19	1.13	1.09	1.03	0.95
Base	1.34	1.28	1.28	1.43	1.47	1.35	1.31	1.25	1.23	1.18	1.12
High	1.34	1.28	1.35	1.56	1.68	1.50	1.41	1.36	1.36	1.32	1.28
Scaling factor (%)											
Low	100%	95%	94%	102%	102%	93%	89%	84%	81%	77%	71%
Base	100%	95%	96%	106%	110%	101%	98%	94%	92%	88%	84%
High	100%	96%	101%	116%	125%	112%	106%	102%	102%	98%	95%

Note: "Total dollars" is, in fact, the total of only a portion of annual Superfund costs, as described in the accompanying text.

Appendix H

Supplementary Tables of Results

Table H-1. Estimated Cost of Fund-Lead Remedial Pipeline Actions at Current NPL Sites: Three Scenarios, FY 2000–FY 2009 (1999$)

Type of action	FY 2000	FY 2001	FY 2002	FY 2003	FY 2004	FY 2005	FY 2006	FY 2007	FY 2008	FY 2009	Total
Low case											
RI/FS	56,004,996	47,768,406	25,449,381	16,922,210	7275,608	3,551,452	1,144,840	261,971	134,163	0	**158,513,027**
RD	44,591,093	59,920,874	59,852,024	42,309,614	29,155,502	22,897,709	18,633,736	10,447,343	8,215,890	7,117,787	**303,141,572**
RA	319,706,941	281,864,667	403,023,463	443,224,456	358,785,884	336,483,785	284,673,605	248,758,240	173,516,468	102,778,851	**2,952,816,360**
LTRA	35,885,000	53,945,000	53,920,000	56,424,800	55,194,200	55,453,800	59,456,124	60,433,744	63,153,840	59,357,900	**553,224,408**
Total	**456,188,030**	**443,498,947**	**542,244,868**	**558,881,080**	**450,411,194**	**418,386,746**	**363,908,305**	**319,901,298**	**245,020,361**	**169,254,538**	**3,967,695,367**
Base case											
RI/FS	56,004,996	47,768,406	25,449,381	16,922,210	7275,608	3,551,452	1,144,840	261,971	134,163	0	**158,513,027**
RD	44,591,093	59,920,874	59,852,024	42,309,614	29,155,502	22,897,709	18,633,736	10,447,343	8,215,890	7,117,787	**303,141,572**
RA	323,716,957	297,944,307	443,974,983	517,425,368	436,091,484	408,895,089	347,901,669	303,714,892	209,890,804	121,826,811	**3,411,382,364**
LTRA	35,885,000	53,945,000	53,920,000	56,424,800	55,194,200	55,453,800	59,456,124	60,433,744	63,153,840	59,357,900	**553,224,408**
Total	**460,198,046**	**459,578,587**	**583,196,388**	**633,081,992**	**527,716,794**	**490,798,050**	**427,136,369**	**374,857,950**	**281,394,697**	**188,302,498**	**4,426,261,371**
High case											
RI/FS	56,819,608	51,876,850	29,965,508	18,144,504	7,435,348	5,600,799	3,194,187	2,311,318	2,183,510	2,049,347	**179,580,980**
RD	38,834,027	67,011,562	90,416,710	61,605,524	36,998,491	30,322,227	23,856,825	9,822,798	7,591,345	5,181,365	**371,640,876**
RA	337,449,277	346,571,795	523,499,143	672,232,344	535,232,028	448,951,105	382,412,565	345,369,900	228,324,516	127,213,563	**3,947,256,235**
LTRA	35,885,000	53,945,000	53,920,000	56,424,800	55,194,200	61,453,800	65,456,124	66,433,744	69,153,840	65,357,900	**583,224,408**
Total	**468,987,912**	**519,405,207**	**697,801,361**	**808,407,172**	**634,860,067**	**546,327,932**	**474,919,702**	**423,937,761**	**307,253,212**	**199,802,176**	**5,081,702,500**

Note: Totals may not add exactly because of rounding.

Table H-2. Estimated Cumulative Number of Construction-Complete Sites on Current NPL, FY 1999–FY 2009

Type of site	FY 1999	FY 2000	FY 2001	FY 2002	FY 2003	FY 2004	FY 2005	FY 2006	FY 2007	FY 2008	FY 2009
Mega sites	25	32	39	44	55	66	72	81	86	87	93
Nonmega sites	623	702	715	754	804	854	922	1,002	1,043	1,071	1,105
Total	**648**	**734**	**754**	**798**	**859**	**920**	**994**	**1,083**	**1,129**	**1,158**	**1,198**

Note: Estimates for FY 2000–FY 2009 are based on RFF remedial pipeline projections; see Appendix G for explanation. Nonfederal sites only.

Table H-3. Estimated Number of Remedial Pipeline Actions Not Yet Completed at Current NPL Sites, FY 1999–FY 2009

Type of action	FY 1999	FY 2000	FY 2001	FY 2002	FY 2003	FY 2004	FY 2005	FY 2006	FY 2007	FY 2008	FY 2009
Fund-Lead											
RI/FS	194	149	90	56	32	14	4	1	1	1	0
RD	263	231	191	138	88	48	37	17	0	5	2
RA	356	305	261	239	204	168	128	101	60	42	32
LTRA	199	192	188	180	178	169	164	159	153	147	109
Total	**1,012**	**877**	**730**	**613**	**502**	**399**	**333**	**278**	**224**	**195**	**143**
PRP-Lead											
RI/FS	210	151	96	67	44	25	14	9	3	1	0
RD	499	428	361	291	195	110	85	54	29	20	9
RA	863	749	650	601	536	446	349	267	103	146	110
PRP LR	410	403	400	392	389	378	368	366	363	355	351
Total	**1,982**	**1,731**	**1,507**	**1,351**	**1,164**	**959**	**816**	**696**	**588**	**522**	**470**

Note: Counts are as of the end of the fiscal year and are inflated slightly (less than 2% annually) because NFA RODs and OU growth are not accounted for in the number of actions.

Table H-4. Estimated Remedial Pipeline Actions Not Yet Completed at Current NPL Sites as Percentage of FY 1999, FY 1999–FY 2009

Type of action	FY 1999	FY 2000	FY 2001	FY 2002	FY 2003	FY 2004	FY 2005	FY 2006	FY 2007	FY 2008	FY 2009
Fund-lead											
RI/FS	100%	77%	46%	29%	16%	7%	2%	1%	1%	1%	0%
RD	100%	88%	73%	52%	33%	18%	14%	6%	4%	2%	1%
RA	100%	86%	73%	67%	57%	47%	36%	28%	17%	12%	9%
LTRA	100%	96%	94%	90%	89%	85%	82%	80%	77%	74%	55%
All actions	**100%**	**87%**	**72%**	**61%**	**50%**	**39%**	**33%**	**27%**	**22%**	**19%**	**14%**
PRP-lead											
RI/FS	100%	72%	46%	32%	21%	12%	7%	4%	1%	0%	0%
RD	100%	86%	72%	58%	39%	22%	17%	11%	6%	4%	2%
RA	100%	87%	75%	70%	62%	52%	40%	31%	22%	17%	13%
PRP LR	100%	98%	98%	96%	95%	92%	90%	89%	89%	87%	86%
All actions	**100%**	**87%**	**76%**	**68%**	**59%**	**48%**	**41%**	**35%**	**30%**	**26%**	**24%**

Note: Based on Table H-3.

Table H-5. Estimated Cost of Fund-Lead Remedial Pipeline Actions at Future NPL Sites: Three Scenarios, FY 2000–FY 2009 (1999$)

Scenario	FY 2000	FY 2001	FY 2002	FY 2003	FY 2004	FY 2005	FY 2006	FY 2007	FY 2008	FY 2009	Total
Low case	6,885,743	18,372,027	28,222,071	34,759,036	38,642,916	44,153,721	57,080,629	77,463,457	99,495,307	114,454,560	**519,529,467**
Base case	6,885,743	20,737,194	35,447,365	47,363,624	56,604,024	67,395,528	90,825,754	131,165,356	179,421,254	218,803,251	**854,649,092**
High case	6,885,743	23,461,596	43,750,362	61,764,384	76,064,828	90,858,914	118,861,707	169,861,959	238,605,426	305,707,463	**1,135,822,381**

Note: Totals may not add exactly because of rounding.

Table H-6. Estimated Cost of Other Site-Specific Activities at Current and Future NPL Sites: Three Scenarios, FY 2000–FY 2009 (1999$)

Activity	FY 2000	FY 2001	FY 2002	FY 2003	FY 2004	FY 2005	FY 2006	FY 2007	FY 2008	FY 2009	Total
Low case											
Oversight of PRP-lead actions	28,869,076	27,676,261	27,255,117	25,878,085	23,143,976	20,601,762	18,912,325	17,101,867	16,270,581	15,427,342	221,136,392
Department of Justice enforcement	25,450,909	24,817,517	23,519,697	23,527,256	22,744,255	21,785,270	24,652,045	27,224,208	24,340,549	22,327,484	240,389,191
EPA enforcement	28,376,001	28,080,400	27,395,601	25,768,763	24,117,768	23,417,551	21,900,335	19,979,232	19,169,273	17,502,738	235,707,663
Response support	13,950,961	13,223,517	13,220,505	12,226,482	10,936,511	9,815,976	9,104,346	8,191,652	7,769,193	7,314,382	105,753,525
Community and state involvement	9,029,367	8,558,549	8,556,600	7,913,246	7,078,349	6,353,114	5,892,532	5,301,817	5,028,391	4,734,028	68,445,993
Five-year reviews	3,625,000	4,050,000	6,200,000	4,975,000	5,300,000	4,650,000	3,700,000	6,325,000	6,950,000	6,775,000	52,550,000
Total	**109,301,314**	**106,406,245**	**106,147,520**	**100,288,832**	**93,320,859**	**86,623,674**	**84,161,583**	**84,123,776**	**79,527,988**	**74,080,974**	**923,982,764**
Base case											
Oversight of PRP-lead actions	28,869,076	27,780,140	27,566,754	26,416,923	23,721,194	21,268,018	19,950,856	18,556,425	18,194,641	17,921,445	230,245,471
Department of Justice enforcement	25,450,909	24,919,151	23,722,965	23,832,158	23,150,791	22,293,440	25,261,849	27,935,646	25,153,621	23,242,190	244,962,722
EPA enforcement	28,376,001	28,540,736	28,316,274	27,149,773	25,959,114	25,719,234	24,662,354	23,201,587	22,851,964	21,645,766	256,422,802
Response support	13,950,961	13,351,535	13,604,559	12,895,188	11,790,467	10,855,182	10,428,204	9,800,162	9,662,355	9,492,196	115,830,809
Community and state involvement	9,029,367	8,641,405	8,805,168	8,346,048	7,631,048	7,025,711	6,749,362	6,342,879	6,253,687	6,143,557	74,968,231
Five-year reviews	3,625,000	4,050,000	6,200,000	4,975,000	5,300,000	4,650,000	3,700,000	6,325,000	6,950,000	6,775,000	52,550,000
Total	**109,301,314**	**107,282,968**	**108,215,720**	**103,615,090**	**97,552,614**	**91,811,585**	**90,752,625**	**92,161,699**	**89,066,268**	**85,220,153**	**974,980,035**

Activity	FY 2000	FY 2001	FY 2002	FY 2003	FY 2004	FY 2005	FY 2006	FY 2007	FY 2008	FY 2009	Total
High case											
Oversight of PRP-lead actions	28,869,076	27,899,370	27,924,444	27,032,520	24,494,911	22,339,555	21,603,449	20,814,380	21,082,265	21,419,292	**243,479,263**
Department of Justice enforcement	25,450,909	25,037,724	23,960,111	24,187,877	23,625,083	22,886,305	25,973,287	28,765,557	26,102,206	24,309,347	**250,298,508**
EPA enforcement	28,376,001	29,077,796	29,390,392	28,760,950	28,107,350	28,404,529	27,884,709	26,961,001	27,148,437	26,479,299	**280,590,464**
Response support	13,950,961	13,497,626	14,042,833	13,654,260	12,752,862	12,020,900	11,914,720	11,607,477	11,790,467	11,941,107	**127,173,213**
Community and state involvement	9,029,367	8,735,959	9,088,828	8,837,336	8,253,931	7,780,189	7,711,457	7,512,613	7,631,048	7,728,545	**82,309,283**
Five-year reviews	3,625,000	4,050,000	6,200,000	4,975,000	5,300,000	4,650,000	3,700,030	6,325,000	6,950,000	6,775,000	**52,550,000**
Total	**109,301,314**	**108,298,475**	**110,606,609**	**107,447,943**	**102,534,138**	**98,081,479**	**98,787,632**	**101,986,128**	**100,704,423**	**98,652,590**	**1,036,400,731**

Note: Totals may not add exactly because of rounding.

Table H-7. Estimated Cost of the Remedial Program: Three Scenarios, FY 2000–FY 2009 (1999$)

Activity	FY 2000	FY 2001	FY 2002	FY 2003	FY 2004	FY 2005	FY 2006	FY 2007	FY 2008	FY 2009	Total
Low case											
Fund-lead actions at current NPL sites	456,188,030	443,498,947	542,244,868	558,881,080	450,411,194	418,386,746	363,908,305	319,901,298	245,020,361	169,254,538	3,967,695,367
Fund-lead actions at future NPL sites	6,885,743	18,372,027	28,222,071	34,759,036	38,642,916	44,153,721	57,080,629	77,463,457	99,495,307	114,454,560	519,529,467
Other site-specific activities at current and future NPL sites	109,301,314	106,406,245	106,147,520	100,288,832	93,320,859	86,623,674	84,161,583	84,123,776	79,527,988	74,080,974	923,982,764
Non-NPL sites	29,399,656	29,399,656	29,399,656	29,399,656	29,399,656	29,399,656	29,399,656	29,399,656	29,399,656	29,399,656	293,996,560
Total	**601,774,742**	**597,676,875**	**706,014,115**	**723,328,605**	**611,774,625**	**578,563,797**	**534,550,173**	**510,888,186**	**453,443,311**	**387,189,728**	**5,705,204,158**
Base case											
Fund-lead actions at current NPL sites	460,198,046	459,578,587	583,196,388	633,081,992	527,716,794	490,798,050	427,136,369	374,857,950	281,394,697	188,302,498	4,426,261,371
Fund-lead actions at future NPL sites	6,885,743	20,737,194	35,447,365	47,363,624	56,604,024	67,395,528	90,825,754	131,165,356	179,421,254	218,803,251	854,649,092
Other site-specific activities at current and future NPL sites	109,301,314	107,282,968	108,215,720	103,615,090	97,552,614	91,811,585	90,752,625	92,161,699	89,066,268	85,220,153	974,980,035
Non-NPL sites	29,399,656	29,399,656	29,399,656	29,399,656	29,399,656	29,399,656	29,399,656	29,399,656	29,399,656	29,399,656	293,996,560
Total	**605,784,758**	**616,998,405**	**756,259,128**	**813,460,362**	**711,273,088**	**679,404,819**	**638,114,403**	**627,584,662**	**579,281,875**	**521,725,558**	**6,549,887,058**

Activity	FY 2000	FY 2001	FY 2002	FY 2003	FY 2004	FY 2005	FY 2006	FY 2007	FY 2008	FY 2009	Total
High case											
Fund-lead actions at current NPL sites	468,987,912	519,405,207	697,801,361	808,407,172	634,860,067	546,327,932	474,919,702	423,937,761	307,253,212	199,802,176	**5,081,702,500**
Fund-lead actions at future NPL sites	6,885,743	23,461,596	43,750,362	61,764,384	76,064,828	90,858,914	118,861,707	169,861,959	238,605,426	305,707,463	**1,135,822,381**
Other site-specific activities at current and future NPL sites	109,301,314	108,298,475	110,606,609	107,447,943	102,534,138	98,081,479	98,787,632	101,986,128	100,704,423	98,652,590	**1,036,400,731**
Non-NPL sites	29,399,656	29,399,656	29,399,656	29,399,656	29,399,656	29,399,656	29,399,656	29,399,656	29,399,656	29,399,656	**293,996,560**
Total	**614,574,624**	**680,564,934**	**881,557,988**	**1,007,019,156**	**842,858,688**	**764,667,980**	**721,968,696**	**725,185,503**	**675,962,717**	**633,561,885**	**7,547,922,171**

Note: Totals may not add exactly because of rounding.

Table H-8. Estimated Total Annual Cost to EPA of Superfund Program: Three Scenarios, FY 2000–FY 2009 (1999$)

Program	FY 2000	FY 2001	FY 2002	FY 2003	FY 2004	FY 2005	FY 2006	FY 2007	FY 2008	FY 2009	Total
Low case											
Removal program	317,845,995	317,845,995	317,845,995	317,845,995	317,845,995	317,845,995	317,845,995	317,845,995	317,845,995	317,845,995	**3,178,459,950**
Remedial program	601,774,742	597,676,875	706,014,115	723,328,605	611,774,625	578,563,797	534,550,173	510,888,186	453,443,311	387,189,728	**5,705,204,158**
Site assessment	57,802,949	57,802,949	57,802,949	57,802,949	57,802,949	57,802,949	57,802,949	57,802,949	57,802,949	57,802,949	**578,029,490**
Program staff, management, and support	304,195,806	296,601,614	293,909,541	278,201,693	259,249,742	240,509,020	224,351,286	211,434,272	206,043,646	194,600,424	**2,509,097,044**
Program administration	108,502,393	107,579,471	116,497,745	116,631,122	105,810,990	101,557,122	96,302,136	93,039,147	88,036,263	81,722,080	**1,015,678,470**
Other programs and agencies	100,500,000	100,500,000	100,500,000	100,500,000	100,500,000	100,500,000	100,500,000	100,500,000	100,500,000	100,500,000	**1,005,000,000**
Total	**1,490,621,885**	**1,478,006,904**	**1,592,570,346**	**1,594,310,364**	**1,452,984,300**	**1,396,778,883**	**1,331,352,539**	**1,291,510,549**	**1,223,672,164**	**1,139,661,176**	**13,991,469,111**
Base case											
Removal program	317,845,995	317,845,995	317,845,995	317,845,995	317,845,995	317,845,995	317,845,995	317,845,995	317,845,995	317,845,995	**3,178,459,950**
Remedial program	605,784,758	616,998,405	756,259,128	813,460,362	711,273,088	679,404,819	638,114,403	627,584,662	579,281,875	521,725,558	**6,549,887,058**
Site assessment	57,802,949	57,802,949	57,802,949	57,802,949	57,802,949	57,802,949	57,802,949	57,802,949	57,802,949	57,802,949	**578,029,490**
Program staff, management, and support	304,195,806	299,524,375	301,158,366	290,090,260	274,688,428	259,497,827	247,979,835	239,702,563	238,951,680	232,148,201	**2,687,937,340**
Program administration	108,836,844	109,426,351	121,276,194	125,114,875	115,363,675	111,509,493	106,860,223	105,071,106	101,209,299	95,999,092	**1,100,667,151**
Other programs and agencies	100,500,000	100,500,000	100,500,000	100,500,000	100,500,000	100,500,000	100,500,000	100,500,000	100,500,000	100,500,000	**1,005,000,000**
Total	**1,494,966,352**	**1,502,098,076**	**1,654,842,632**	**1,704,814,441**	**1,577,474,135**	**1,526,561,083**	**1,469,103,405**	**1,448,507,274**	**1,395,591,797**	**1,326,021,795**	**15,099,980,990**

Program	FY 2000	FY 2001	FY 2002	FY 2003	FY 2004	FY 2005	FY 2006	FY 2007	FY 2008	FY 2009	Total
High case											
Removal program	317,845,995	317,845,995	317,845,995	317,845,995	317,845,995	317,845,995	317,845,995	317,845,995	317,845,995	317,845,995	**3,178,459,950**
Remedial program	614,574,624	680,564,934	881,557,988	1,007,019,156	842,858,688	764,667,980	721,968,696	725,185,503	675,962,717	633,561,885	**7,547,922,171**
Site assessment	57,802,949	57,802,949	57,802,949	57,802,949	57,802,949	57,802,949	57,802,949	57,802,949	57,802,949	57,802,949	**578,029,490**
Program staff, management, and support	304,195,806	302,898,493	309,508,016	303,729,123	292,328,767	281,139,642	274,910,862	271,922,802	276,461,131	274,946,864	**2,892,041,507**
Program administration	109,569,954	114,999,688	132,403,452	142,366,644	127,770,592	120,376,893	116,041,480	115,830,241	112,323,063	108,808,246	**1,200,490,254**
Other programs and agencies	100,500,000	100,500,000	100,500,000	100,500,000	100,500,000	100,500,000	100,500,000	100,500,000	100,500,000	100,500,000	**1,005,000,000**
Total	**1,504,489,328**	**1,574,612,059**	**1,799,618,401**	**1,929,263,867**	**1,739,106,992**	**1,642,333,460**	**1,589,069,983**	**1,589,087,490**	**1,540,895,855**	**1,493,465,939**	**16,401,943,373**

Note: Totals may not add exactly because of rounding.

Table H-9. Estimated Total Annual Cost to EPA of Superfund Program: Three Scenarios, FY 2000–FY 2009 (Inflation-Adjusted Dollars)

Program	FY 2000	FY 2001	FY 2002	FY 2003	FY 2004	FY 2005	FY 2006	FY 2007	FY 2008	FY 2009	Total
Low case											
Removal program	323,885,069	331,334,425	338,623,783	345,734,882	352,303,845	358,997,618	365,818,573	372,769,126	379,851,739	387,068,922	**3,556,387,983**
Remedial program	613,208,462	623,040,488	752,166,692	786,795,914	678,097,431	653,470,638	615,229,968	599,168,609	541,901,528	471,514,863	**6,334,594,592**
Site assessment	58,901,205	60,255,933	61,581,563	62,874,776	64,069,397	65,286,715	66,527,163	67,791,179	69,079,211	70,391,716	**646,758,859**
Program staff, management, and support	309,975,526	309,188,497	313,122,588	302,612,055	287,355,141	271,647,800	258,212,683	247,969,677	246,238,865	236,981,990	**2,783,304,823**
Program administration	110,563,938	112,144,821	124,113,274	126,864,733	117,282,015	114,705,755	110,837,042	109,116,119	105,210,474	99,520,139	**1,130,358,310**
Other programs and agencies	102,409,500	104,764,919	107,069,747	109,318,211	111,395,257	113,511,767	115,668,491	117,866,192	120,105,650	122,387,657	**1,124,497,391**
Total	**1,518,943,701**	**1,540,729,083**	**1,696,677,647**	**1,734,200,571**	**1,610,503,086**	**1,577,620,294**	**1,532,293,921**	**1,514,680,901**	**1,462,387,468**	**1,387,865,287**	**15,575,901,959**
Base case											
Removal program	323,885,069	331,334,425	338,623,783	345,734,882	352,303,845	358,997,618	365,818,573	372,769,126	379,851,739	387,068,922	**3,556,387,983**
Remedial program	617,294,669	643,181,967	805,696,252	884,836,137	788,382,574	767,367,580	734,425,175	736,029,994	692,288,816	635,350,933	**7,304,854,095**
Site assessment	58,901,205	60,255,933	61,581,563	62,874,776	64,069,397	65,286,715	66,527,163	67,791,179	69,079,211	70,391,716	**646,758,859**
Program staff, management, and support	309,975,526	312,235,291	320,845,273	315,543,765	304,467,544	293,095,094	285,407,495	281,122,670	285,566,634	282,707,208	**2,990,966,501**
Program administration	110,904,744	114,070,077	129,204,094	136,092,879	127,870,311	125,946,663	122,988,664	123,227,175	120,953,319	116,906,508	**1,228,164,433**
Other programs and agencies	102,409,500	104,764,919	107,069,747	109,318,211	111,395,257	113,511,767	115,668,491	117,866,192	120,105,650	122,387,657	**1,124,497,391**
Total	**1,523,370,713**	**1,565,842,612**	**1,763,020,711**	**1,854,400,651**	**1,748,488,928**	**1,724,205,437**	**1,690,835,561**	**1,698,806,336**	**1,667,845,370**	**1,614,812,944**	**16,851,629,263**

Program	FY 2000	FY 2001	FY 2002	FY 2003	FY 2004	FY 2005	FY 2006	FY 2007	FY 2008	FY 2009	Total
High case											
Removal program	323,885,069	331,334,425	338,623,783	345,734,882	352,303,845	358,997,618	365,818,573	372,769,126	379,851,739	387,068,922	**3,556,387,983**
Remedial program	626,251,542	709,446,068	939,185,975	1,095,378,437	934,233,438	863,669,789	830,935,618	850,496,058	807,830,262	771,543,828	**8,428,971,017**
Site assessment	58,901,205	60,255,933	61,581,563	62,874,776	64,069,397	65,286,715	66,527,163	67,791,179	69,079,211	70,391,716	**646,758,859**
Program staff, management, and support	309,975,526	315,752,597	329,740,745	330,379,348	324,020,281	317,538,882	316,403,229	318,910,500	330,393,471	334,826,891	**3,227,941,471**
Program administration	111,651,783	119,879,929	141,058,747	154,858,377	141,622,269	135,962,128	133,555,661	135,845,467	134,235,169	132,505,337	**1,341,174,858**
Other programs and agencies	102,409,500	104,764,919	107,069,747	109,318,211	111,395,257	113,511,767	115,668,491	117,866,192	120,105,650	122,387,657	**1,124,497,391**
Total	**1,533,074,625**	**1,641,433,871**	**1,917,260,560**	**2,098,544,033**	**1,927,644,488**	**1,854,966,900**	**1,828,908,724**	**1,863,678,523**	**1,841,495,503**	**1,818,724,352**	**18,325,731,579**

Note: Totals may not add exactly because of rounding.

Abbreviations and Acronyms

ARAR	applicable or relevant and appropriate requirement
ASTSWMO	Association of State and Territorial Solid Waste Management Officials
CBO	Congressional Budget Office
CEPPO	(U.S. EPA) Chemical Emergency Preparedness and Prevention Office
CERCLA	Comprehensive Environmental Response, Compensation, and Liability Act of 1980
CERCLIS	Comprehensive Environmental Response, Compensation, and Liability Information System
DOJ	(U.S.) Department of Justice
EPA	(U.S.) Environmental Protection Agency
FTE	full-time equivalent
FY	fiscal year
GAO	(U.S.) General Accounting Office
HRS	hazard ranking system
IFMS	Integrated Financial Management System
LTRA	long-term response action
NCP	National Contingency Plan
NFA	no further action
NFRAP	no further remedial action planned
NIEHS	National Institute for Environmental Health Sciences
NPL	National Priorities List
NRC	National Response Center
O&M	operation and maintenance

OARM	(U.S. EPA) Office of Administration and Resources Management
OCFO	(U.S. EPA) Office of the Chief Financial Officer
OECA	(U.S. EPA) Office of Enforcement and Compliance Assurance
OIG	(U.S. EPA) Office of the Inspector General
OSC	on-scene coordinator
OSWER	(U.S. EPA) Office of Solid Waste and Emergency Response
OU	operable unit
PA	preliminary assessment
PRP	potentially responsible party
PRP LR	long-term response action conducted by a potentially responsible party(ies)
RA	remedial action
RCRA	Resource Conservation and Recovery Act
RD	remedial design
RFF	Resources for the Future
RI/FS	remedial investigation/feasibility study
ROD	record of decision
RPM	remedial project manager
SACM	Superfund accelerated cleanup model
SARA	Superfund Amendments and Reauthorization Act
SI	site inspection
TIO	(U.S. EPA) Technology Innovation Office
TMDL	total maximum daily load

Glossary
of Terms

construction-complete: a site at which the physical construction of all cleanup actions is complete, all immediate threats have been addressed, and all long-term threats are under control. Construction-complete sites can still have one (the last) remedial action under way.

deleted NPL site: a National Priorities List site at which EPA has determined, with state concurrence, that no further response is required to protect human health or the environment. After EPA approves a "close-out" report establishing that all response actions have been taken or that no action is required, EPA publishes a deletion notice in the *Federal Register*.

emergency response: a removal action that, based on the lead agency's evaluation of the release or threat of release of hazardous substances, must begin within hours.

extramural cost: expenditures made by EPA that are "external" to the agency, including contracts, interagency agreements, and cooperative agreements with states.

final NPL site: a site, usually with a hazard ranking system score of 28.5 or more, that has been added to the National Priorities List through the issuance of a final rule in the *Federal Register*. EPA can use Trust Fund monies to pay for long-term remedial actions only at final NPL sites.

fiscal year (FY): a yearly accounting period, regardless of its relationship to a calendar year. The fiscal year for the federal government runs from October 1 to September 30 of the next year, and is designated by the calendar year in which it ends.

Fund-lead action: an action financed, in whole or in part, and conducted by EPA (often by the U.S. Army Corps of Engineers, contractors, or state agencies).

hazard ranking system (HRS): the system EPA uses to score potential risks to human health and the environment from releases or threatened releases of hazardous substances at a site. In general, a site must score at least 28.5 to be placed on the National Priorities List.

intramural cost: expenditures made by EPA that are "internal" to the Agency, including payroll, travel, and supplies.

long remedial action: a remedial action with an expected duration of 10 years or longer.

long-term response action (LTRA): Fund-financed operation of ground-water and surface water restoration measures, including monitored natural attenuation, for the first 10 years of operation.

mega site: a site with actual or expected total removal and remedial action costs of $50 million or more.

nonmega site: a site with actual or expected total removal and remedial action costs of less than $50 million.

nonteenager site: a site listed on the NPL that was proposed for listing prior to FY 1987 and *was* construction-complete at the end of FY 1999; or a site proposed to the NPL for listing in FY 1987 or later.

non-time-critical removal action: a removal action that, based on a site evaluation, the lead agency determines does not need to be initiated within the next six months.

NPL equivalent site: a site not listed as final on the National Priorities List at which responsible parties perform remedial action under EPA enforcement authority and with EPA oversight in accordance with the cleanup provisions set out in the National Contingency Plan.

operable unit (OU): a distinct project of the overall site cleanup. Sites can be divided into operable units based on the media to be addressed (such as groundwater or contaminated soil), geographic area, or other measures.

operation and maintenance (O&M): activities required to maintain the effectiveness and integrity of a remedy, including groundwater pumping and treating, measures to restore groundwater or surface water, and maintenance of landfill caps.

OU growth: the subdivision of an operable unit into two or more operable units.

OU lag: the time between the start of remedial response actions at multiple operable units at a site.

potentially responsible party (PRP): an individual, business, or other organization that is potentially liable for cleaning up a site. The four types of responsible parties include a site's present owner(s) and operator(s), its previous owner(s) and operator(s) during the time when it received hazardous substances, the generators of such substances, and any waste transporters responsible for choosing the site.

preliminary assessment: the first stage of EPA's screening process for investigating suspected contaminated sites, generally involving review of available documents.

proposed NPL site: a site that has been proposed for listing on the National Priorities List through the issuance of a proposed rule in the *Federal Register*. EPA then accepts public comments on the site, responds to the comments, and places on the NPL those sites that continue to meet the requirements for listing.

PRP-lead action: an action conducted and financed by a potentially responsible party or parties. A portion of the response action may be paid for from the Trust Fund through a preauthorized reimbursement under section 106(b) of CERCLA, a practice referred to as preauthorized mixed funding.

PRP LR: groundwater and surface water restoration measures, including monitored natural attenuation, conducted by a potentially responsible party. PRP LR is the equivalent of LTRA, except that LTRA is conducted by EPA.

record of decision (ROD): the public document in which EPA identifies the cleanup alternative to be used at an operable unit of a site.

remedial action (RA): the actual construction or implementation of a remedy at a site or portion thereof.

remedial design (RD): the engineering plan for cleaning up a site or portion thereof. The actual remedial design document includes the technical drawings and specifications that will guide implementation of the remedy, referred to as the remedial action.

remedial investigation/feasibility study (RI/FS): site studies that involve gathering data to determine the type and extent of contamination at a site (or portion thereof), establishing cleanup criteria, and analyzing the feasibility and costs of alternative cleanup methods. The study can be conducted by EPA, contractors, state agencies, or potentially responsible parties.

site inspection: the second stage of EPA's process for screening a contaminated site to determine whether it warrants inclusion on the National Priorities List. The site inspection involves collecting and analyzing samples of soil and water.

teenager site: a site listed on the National Priorities List that was proposed for listing prior to FY 1987 and that, as of the end of FY 1999, was not construction-complete. In other words, the site has been on the NPL and is still not construction-complete after at least 13 years, making it a "teen-ager" site.

time-critical removal action: a removal action that, based on a site evalua-tion, the lead agency determines must be initiated within six months.

Trust Fund: the Trust Fund created by Congress to finance EPA's implemen-tation of the Superfund program, officially called the Hazardous Substance Superfund.

Notes

Chapter One: Introduction

1. Public Law No. 96-510, 94 Stat. 2767 (1980) (codified as amended at 42 U.S.C. secs. 9601–9675 (1988 & Supp. IV 1992)).

2. U.S. General Accounting Office (GAO), *Major Management Challenges and Program Risks: Environmental Protection* Agency, GAO-01-257, 2001, 36; Robert Nakamura and Thomas Church, *Reinventing Superfund: An Assessment of EPA's Administrative Reforms*, prepared for the National Academy of Public Administration, 2000; Superfund Settlements Project, *The Superfund Program at Its 20th Anniversary: Achievements of the Past, Challenges for the Future*, December 11, 2000.

3. U.S. Environmental Protection Agency (EPA), *Close Out Procedures for National Priorities List Sites*, EPA-540-R-98-016, 2000, 3–1.

4. Congressional Budget Office (CBO), *Budget Options*, 2001, p. 217.

5. Departments of Veterans Affairs and Housing and Urban Development, and Independent Agencies Appropriations Act, 2000, Conference Report 106-379.

6. See Katherine N. Probst, Don Fullerton, Robert E. Litan, and Paul R. Portney, *Footing the Bill for Superfund Cleanups: Who Pays and How?* (Washington, DC: The Brookings Institution Press and Resources for the Future, 1995), for an earlier RFF effort to estimate the cost of Superfund to potentially responsible parties.

7. The one exception is site-specific special accounts, which are funds obtained through settlements with PRPs and designated for future costs at specific sites. Expenditures from these accounts are included in our analyses.

8. See, for instance, U.S. GAO, *Superfund: Systems Enhancements Could Improve Efficiency of Cost Recovery*, GAO/AIMD-95-177, 1995.

9. The term *mega site* was first used in a 1994 CBO report estimating future Superfund costs. CBO, *The Total Costs of Cleaning Up Nonfederal Superfund Sites*, 1994.

10. Because this research was funded under a contract with EPA, the Paperwork Reduction Act limited us to interviewing staff in nine states.

11. Transcript of *Marketplace*, October 12, 2000, Minnesota Public Radio, prepared by Burrelle's Information Services, p. 7.

12. For FY 2001, Congress appropriated $1.27 billion to EPA for the Superfund program.

Chapter Two: The Removal Program

1. Comprehensive Environmental Response, Compensation, and Liability Act (CERCLA) sec. 104(a)(1).

2. U.S. Environmental Protection Agency (EPA), *Superfund Removal Procedures Response Management: Removal Action Start-Up to Close-Out*, EPA 540-R-96-039, 1996, p. 1.

3. CERCLA sec. 104(c)(1).

4. Based on EPA analysis of CERCLIS and IFMS data provided to RFF, January 2001. A removal action was considered to have costs of more than $2 million if there was a flag in CERLCIS to that effect or if, according to IFMS data, total expenditures (in nominal dollars) exceeded $2 million. As of January 19, 2001, 304 removal actions (of a total of 3,621 Fund-lead removal actions for which there were complete data) undertaken at nonfederal facility sites met one of these two criteria.

5. This figure may, in fact, be an underestimate, since the NPL status of a site recorded in CERCLIS reflects the status of the site when the data were extracted from the database for the RFF study, not the status of the site at the time of the action.

6. Presidential Decision Directive (PDD) 39, "U.S. Policy on Counter-Terrorism," June 21, 1995; PDD-62, "Protection against Unconventional Threats to the Homeland and Americans Overseas," May 22, 1998; PDD-63, "Critical Infrastructure Protection," May 22, 1998; PDD-67, "Enduring Constitutional Government and Continuity of Government," October 21, 1998; and U.S. EPA, "Recommendations for Regional Implementation of EPA's FY 2000 Counter-Terrorism Program," memorandum from Timothy Fields, Jr., Assistant Administrator for the Office of Solid Waste and Emergency Response to Superfund National Policy Managers, Regions 1–10, November 8, 1999.

7. U.S. EPA, *Superfund Removal Procedures*, p. 2.

8. In FY 1999, EPA had 186.5 full-time equivalents (FTE) working as on-scene coordinators, spread across all 10 EPA regional offices. This number of FTE was determined by applying ratios for FY 2000 to the total number of FTE in FY 1999.

9. "Removals historical summary data" provided to RFF by EPA, March 2000. (See Appendix C for details.)

10. Ibid.

11. Association of State and Territorial Solid Waste Management Officials (ASTSWMO) and U.S. EPA, *Steering Committee Evaluation of State-Lead Time-Critical Removal Pilots—Final Report*, Washington, DC, 2000. The report states, "The Removal Action Focus Group recommends that EPA maintain their current response capabilities

and utilize the States' removal capacities to enhance national removal actions to bring about more and earlier risk reduction at more sites" (p. ii).

12. This total includes only payroll costs directly attributable to the removal program. There are some additional costs for benefits, equipment and supplies, travel, and training, which are not included because they were not easily separable (but likely total between $2 million and $3 million) as well as other costs that are not readily disaggregated from general program management and administrative expenses. These costs are discussed in Chapter 6.

13. Expenditure data from IFMS provided to RFF by EPA, January 2001.

14. "Release notifications data" provided to RFF by EPA, April 2000.

15. These average costs are based on a total of 1,108 sites at which there was only a single Fund-lead removal action taken: 364 emergency responses, 729 time-critical removal actions, and 15 non-time-critical removal actions.

16. All estimates of removal action costs by type are based on an analysis of IFMS expenditure data conducted by EPA and provided to RFF, January 2001.

17. Personal communication between Craig Olewiler and Katherine N. Probst, January 9, 2001.

18. Disaster Relief and Emergency Assistance Act (Stafford Act), Public Law No. 93-288 (1988) (codified as amended at 42 U.S.C. secs. 5121–5204).

19. Delays typically occur when EPA does not provide the needed documentation in a timely manner.

Chapter Three: The Remedial Program and the Current National Priorities List

1. Comprehensive Environmental Response, Compensation, and Liability Information System (CERCLIS) and archive data as of October 18, 2000.

2. Statement of Carol M. Browner, Administrator, U.S. Environmental Protection Agency (EPA), before the Subcommittee on Finance and Hazardous Materials, House Committee on Commerce, March 5, 1998.

3. Comprehensive Environmental Response, Compensation, and Liability Act (CERCLA) sec. 107(a)(1)–(4).

4. Under CERCLA section 107(a), when EPA formally notifies a party of its potential liability at a Superfund site, it is considered a PRP.

5. CERCLA sec. 106(b)(1).

6. CERCLA sec. 107(c)(3).

7. Data provided to RFF by EPA, October 2000.

8. CERCLA sec. 104(c)(1).

9. 40 C.F.R. Part 300.430(f)(1)(i)(B).

10. There were an additional 173 federal facility sites on the NPL at the end of FY 1999.

11. U.S. EPA, *Close Out Procedures for National Priorities List Sites*, EPA 540-R-98-016, 2000, 3–1.

12. These 707 sites do not include 4 sites proposed to the NPL in FY 1982 and 1 site proposed to the NPL in FY 1983, all of which were shortly thereafter deleted from the NPL without ever having been listed as final.

13. The Congressional Budget Office (CBO) first used this approach in a 1994 report that estimated the future costs of the Superfund program. CBO, *The Total Costs of Cleaning Up Nonfederal Superfund Sites,* 1994.

14. E.W. Colglazier, T. Cox, and K. Davis, *Estimating Resource Requirements for NPL Sites* (Knoxville, TN: University of Tennessee, Waste Management Research and Education Institute, 1991); Katherine N. Probst, Don Fullerton, Robert E. Litan, and Paul R. Portney, *Footing the Bill for Superfund Cleanups: Who Pays and How?* (Washington, DC: The Brookings Institution and Resources for the Future, 1995).

15. 40 C.F.R. Part 300.425(b)(1).

16. Memo on NPL equivalent sites provided to RFF by EPA, September 2000.

17. Ibid.

18. Data provided to RFF by EPA, October 2000.

19. 40 C.F.R. Part 300.510(b)(1) and 40 C.F.R. Part 300.510(c)(1).

20. CBO, *The Total Costs of Cleaning Up;* Colglazier, et al., *Estimating Resource Requirements.*

21. The U.S. General Accounting Office (GAO) used IFMS data in a May 1999 report that characterized EPA expenditures on a broad level (e.g., contractor cleanup work, site-specific support and nonsite-specific support), but it did not use the data to review the costs of individual cleanup actions. U.S. GAO, *Superfund: EPA Can Improve Its Monitoring of Superfund Expenditures,* GAO/RCED-99-139, 1999.

22. Presumptive remedies are "preferred technologies or response actions for common categories of [Superfund] sites based on historical patterns of remedy selection and EPA's scientific and engineering evaluation of performance data on technology innovation." U.S. EPA, *Presumptive Remedies: Policy and Procedures,* EPA 540-F-93-047, 1993.

23. Statement of Timothy Fields, Jr., Assistant Administrator, U.S. EPA, before the Subcommittee on Superfund, Waste Control and Risk Assessment, Senate Committee on Environment and Public Works, March 21, 2000.

24. Statement of Carol M. Browner, Administrator, U.S. EPA, before the Subcommittee on Water Resources and Environment, House Committee on Transportation and Infrastructure, May 12, 1999.

25. Robert Nakamura and Thomas Church, *Reinventing Superfund: An Assessment of EPA's Administrative Reforms,* prepared for the National Academy of Public Administration, 2000, p. 46–47.

26. GAO used this approach in its recent study estimating Superfund costs. U.S. GAO, *Superfund: Information on the Program's Funding and Status,* GAO/RCED-00-25, 1999.

27. Statement of Governor Christine Todd Whitman, Administrator, U.S. EPA, before the Subcommittee on Water Resources and Environment, House Committee on Transportation and Infrastructure, May 2, 2001.

Chapter Four: Post-Construction Activities

1. *Federal Register* 55:46, March 8, 1990.

2. Departments of Veterans Affairs and Housing and Urban Development, and Independent Agencies Appropriations Act, 2000, Conference Report 106-379.

3. U.S. Environmental Protection Agency (EPA), *Close Out Procedures for National Priorities List Sites*, EPA 540-R-98-016, 2000, pp. 2–3.

4. 40 C.F.R. Part 300.435(f)(3).

5. Information provided to RFF by EPA, April 2000. EPA has also considered the number of sites that have required LTRA. These studies concluded that 44% to 46% of NPL sites included at least one remedy to restore groundwater.

6. In its 1994 study, the Congressional Budget Office (CBO) assumed that 47% of remedial actions involved contaminated groundwater. CBO, *The Total Costs of Cleaning Up Nonfederal Superfund Sites*, 1994, p. 25, footnote 17.

7. U.S. EPA, *Estimated O&M Costs for RODs: Historical Trends and Projected Costs through FY 2040*, 1995, pp. 3–5; U.S. General Accounting Office (GAO), *Superfund: Operations and Maintenance Activities Will Require Billions of Dollars*, GAO/RCED-95-259, 1995, p. 20.

8. U.S. EPA, *Groundwater Cleanup: Overview of Operating Experience at 28 Sites*, EPA 542-R-99-006, 1999.

9. U.S. EPA, *Superfund Reform Strategy Pump and Treat Optimization Implementation Plan*, OSWER 9283.1-13, 2000, p. 1.

10. Statement of Robert W. Varney, Commissioner, New Hampshire Department of Environmental Services, on Behalf of the Environmental Council of the States, before the Subcommittee on Superfund, Waste Control and Risk Assessment, Senate Committee on Environment and Public Works, March 21, 2000; statement of Claudia Kerbawy, Association of State and Territorial Solid Waste Management Officials, before the Senate Committee on Environment and Public Works, May 25, 1999.

11. GAO reached a similar conclusion in a 1993 report. U.S. GAO, *Superfund: Cleanups Nearing Completion Indicate Future Challenges*, GAO/RCED-93-188, 1993, p. 5.

12. Comprehensive Environmental Response, Compensation, and Liability Act (CERCLA) sec. 121(c).

13. Robert Nakamura and Thomas Church, *Reinventing Superfund: An Assessment of EPA's Administrative Reforms,* prepared for the National Academy of Public Administration, 2000, p. 57.

14. U.S. EPA, *Structure and Components of Five-Year Reviews*, May 23, 1991, pp. 2–3.

15. Ibid., Attachment, p. 2.

16. Ibid., Attachment, pp. 2–3.

17. Ibid., Attachment, p. 3.

18. 40 C.F.R. Part 300.430(f)(4)(ii).

19. U.S. EPA, *Structure and Components*, pp. 2–3.

20. U.S. EPA, *Supplemental Five-Year Review Guidance*, July 26, 1994, p. 5; U.S. EPA, *Second Supplemental Five-Year Review Guidance*, December 21, 1995, Attachment, p. 3.

21. See, for instance, Joel S. Hirschhorn, "EPA's Five-Year Review of Superfund Sites Needs Higher Priority," *Environment Reporter*, February 26, 1999; and Nakamura and Church, *Reinventing Superfund*, pp. 60–62.

22. U.S. GAO, *Superfund: Operations and Maintenance Activities*.

23. U.S. EPA, Office of the Inspector General (OIG), *Backlog Warrants Higher Priority for Five-Year Reviews*, 1995, p. 4.

24. U.S. EPA, OIG, *Superfund Backlog of Five-Year Review Reports Increased Nearly Threefold*, 1999, pp. 6–7, 12–20.

25. Based on data from RFF action-level dataset and five-year review completion data provided to RFF by EPA, March 2000 and January 2001.

26. U.S. EPA, OIG, *Superfund Backlog*, p. 10.

27. Ibid., pp. 12–13.

28. Hirschhorn, "EPA's Five-Year Review," pp. 2, 151; and Nakamura and Church, *Reinventing Superfund*, p. 61.

29. U.S. EPA, *Five-Year Review for the CTS Printex Superfund Site*, November 10, 1999, p. 4.

30. U.S. EPA, *Second Supplemental Five-Year Review Guidance*, Attachment, p. 3.

31. U.S. EPA, *Five-Year Review Report for the Denver Radium Site (Shattuck Chemical) Operable Unit #8*, December 20, 1999, pp. 1–3.

32. U.S. EPA, *Record of Decision Amendment, Denver Radium Site, Operable Unit 8*, June 16, 2000, p. 2.

33. Ibid., p. 3.

34. Information provided to RFF by EPA, February 2000.

35. Ibid.

36. Based on IFMS expenditure data, extramural dollars only.

37. Information provided to RFF by EPA, November 2000.

38. Ibid.

39. Robert Hersh, Katherine N. Probst, Kris Wernstedt, and Jan Mazurek, *Linking Land Use and Superfund Cleanups: Uncharted Territory* (Washington, DC: RFF, 1997), p. 65.

Chapter Five: The Size and Composition of the Future National Priorities List

1. Congressional Budget Office (CBO), *The Total Costs of Cleaning Up Nonfederal Superfund Sites*, 1994; U.S. Congress, Office of Technology Assessment (OTA), *Coming Clean: Superfund Problems Can Be Solved*, OTA-ITE-433, 1989. EPA estimates were described in U.S. General Accounting Office (GAO), *Superfund: Estimates of Number of Future Sites Vary*, GAO/RCED-95-18, 1994.

2. Statement of Claudia Kerbawy, Chair, Federal Superfund Focus Group, Association of State and Territorial Solid Waste Management Officials (ASTSWMO), before the Subcommittee on Finance and Hazardous Materials, House Committee on Commerce, March 23, 1999, p. 147.

3. ASTSWMO, *State Cleanup Accomplishments for the Period 1993–1997*, 1998, p. ES-8.

4. U.S. EPA, *Improving Site Assessment: Pre-CERCLIS Screening Assessments*, EPA 540-F-98-039, 1999.

5. Information provided to RFF by EPA, August 2000.

6. Comprehensive Environmental Response, Compensation, and Liability Information System (CERCLIS) data as of October 18, 2000, provided to RFF by EPA.

7. Comprehensive Environmental Response, Compensation, and Liability Act (CERCLA) sec. 105(a)(8)(B); U.S. Congress, OTA, *Coming Clean*, p. 195.

8. U.S. EPA, *Superfund Fact Sheet: Identifying Sites*, OSWER 9230.0-05FS, 1992.

9. U.S. Congress, OTA, *Coming Clean*, p. 196.

10. U.S. GAO, *Hazardous Waste: Unaddressed Risks at Many Potential Superfund Sites*, GAO/RCED-99-8, 1998, p. 27.

11. Ibid., p. 21.

12. CBO, *The Total Costs of Cleaning Up*; EPA estimates were described in U.S. GAO, *Superfund: Estimates*.

13. Prepared statement of Hon. John D. Dingell (D-MI) before the Subcommittee on Finance and Hazardous Materials, House Committee on Commerce, March 23, 1999.

14. U.S. GAO, *Hazardous Waste*, p. 22.

15. EPA has been tracking the status of the 3,036 sites identified in the GAO report. As of July 2000, 48% of the 109 sites proposed to the NPL since March 1998 had come from the GAO universe of sites. The balance of recent proposals to the NPL, 52%, did not come from the GAO universe and included sites that had been added to CERCLIS as well as sites that, at the time of the GAO investigation, were in the earlier stages of the CERCLIS pipeline. Since the GAO study was published, EPA regions and states have been negotiating work-share arrangements to discuss responsibility for additional site assessment at the sites identified by GAO. There has been some progress. According to EPA data as of July 2000, of the 3,036 sites identified by GAO, 131 sites have been designated "no further remedial action planned" (NFRAP), and 333 sites have been archived.

16. Interview with EPA Region 4, May 25, 2000.

17. CERCLIS data provided to RFF by EPA.

18. Environmental Law Institute (ELI), *An Analysis of State Superfund Programs: 50-State Study, Update* (Washington, DC: ELI, 1998).

19. U.S. GAO, *Superfund: Extent of Nation's Potential Hazardous Waste Problem Still Unknown*, GAO/RCED-88-44, 1987; Robert A. Simons, *Turning Brownfields into Greenbacks: Developing and Financing Environmentally Contaminated Urban Real Estate* (Washington, DC: Urban Land Institute, 1998).

20. CERCLA 105(a)(1).

21. Approximately half the states in the country have such laws on their books; see ELI, *An Analysis of State Superfund Programs*.

22. Statement of Timothy Fields, Jr., Assistant Administrator, U.S. EPA, before the Subcommittee on Finance and Hazardous Materials, House Committee on Commerce, March 23, 1999.

23. ASTSWMO, *State Cleanup Accomplishments*.

24. RFF IFMS dataset.

25. CBO, *The Total Costs of Cleaning Up*, pp. 20–21.

26. Interview with EPA Region 2, May 22, 2000.

27. Based on RFF interviews with EPA and state hazardous waste officials, conducted from May through July 2000.

28. Ibid.

29. See National Research Council (NRC), *Contaminated Sediments in Ports and Waterways: Cleanup Strategies and Technologies* (Washington, DC: National Academy Press, 1997); U.S. EPA, *The Incidence and Severity of Sediment Contamination in Surface Waters of the United States: Volume 1: National Sediment Quality Survey*, EPA 823-R-97-006, 1997; U.S. EPA, *EPA's Contaminated Sediment Management Strategy*, EPA-823-R-98-001, 1998; National Oceanic and Atmospheric Administration, *Inventory of Chemical Concentrations in Coastal and Estuarine Sediments*, 1994.

30. U.S. EPA, *EPA's Contaminated Sediment Management Strategy*.

31. U.S. EPA, *The Incidence and Severity of Sediment Contamination*.

32. RFF interview with Superfund regional division director, May 2000.

33. U.S. EPA, *National Hardrock Mining Framework*, EPA 833-B-97-003, 1997.

34. NRC, *Hardrock Mining on Federal Lands* (Washington, DC: National Academy Press, 1999), p. 154.

35. U.S. GAO, *EPA's Expenditures to Clean Up the Bunker Hill Superfund Site*, GAO-01-431R, 2001, p. 4.

36. RFF interviews with EPA regional and state hazardous waste officials, conducted from May through July 2000.

37. Ibid.

38. James R. Kuipers, *Hardrock Reclamation Bonding Practices in the Western United States*, prepared for the National Wildlife Federation, 2000, p. 1.

39. Kuipers, *Hardrock Reclamation Bonding Practices*, p. 2.

40. NRC, *Hardrock Mining*, p. 201.

41. ASTSWMO, *State Cleanup Accomplishments*.

42. Ibid.

43. U.S. GAO, *Community Development: Reuse of Urban Industrial Sites*, GAO/RCED-95-172, 1995; U.S. GAO, *Hazardous Waste Sites: State Cleanup Practices* GAO/RCED-99-39, 1998.

44. ELI, *An Analysis of State Superfund Programs*, p. 76.

45. RFF interview with New York DEC officials, February 28, 2001.

46. Ibid.

47. RFF interview with official from Pennsylvania Department of Environmental Protection, February 27, 2001.

48. RFF state interviews, July 2000.

49. Statement of Peter F. Guerrero, GAO, Director, Environmental Protection Issues, Subcommittee on Finance and Hazardous Materials, House Committee on Commerce, March 23, 1999, p. 134.

50. U.S. EPA, Memorandum from Elliott P. Laws, Assistant Administrator, OSWER, to Regional Administrators: Coordinating with States on National Priorities List Decisions, November 14, 1996.

51. Data provided to RFF by EPA, October 2000. Based on actual RA starts from FY 1996 through FY 2000.

Chapter Six: Superfund Support Activities and Programs

1. Based on payroll data provided to RFF by EPA, August 2000.

2. Information provided to RFF by EPA, September 2000.

3. William A. Suk, *Superfund Basic Research Program, Program Overview 1999*, prepared for National Institute for Environmental Health Sciences, 1999.

4. Information provided to RFF by EPA, February 2000.

5. Information provided to RFF by Department of Justice, October 2000.

Chapter Seven: The Future Cost of Superfund

1. Departments of Veterans Affairs and Housing and Urban Development, and Independent Agencies Appropriations Act, 2000, Conference Report 106-379.

2. Congressional Budget Office (CBO), *The Total Costs of Cleaning Up Nonfederal Superfund Sites*, 1994.

3. U.S. General Accounting Office (GAO), *Superfund: Information on the Program's Funding and Status*, GAO/RCED-00-25, 1995, p. 12.

4. Statement of Timothy Fields, Jr., Assistant Administrator, U.S. EPA, before the Subcommittee on Superfund, Waste Control, and Risk Assessment, Senate Committee on Environment and Public Works, March 21, 2000.

5. U.S. EPA, *Strategic Plan*, EPA 190-R-00-002, 2000, p. 35.

6. Statement of Governor Christine Todd Whitman, Administrator, U.S. EPA, before the Subcommittee on Water Resources and Environment, House Committee on Transportation and Infrastructure, May 2, 2001.

7. This information is based on our interviews with EPA regional division directors, as discussed in Chapter 5.

8. The average number of operable units for mega sites and nonmega sites is based on the average for these sites calculated from sites on the current NPL at the end of FY 1999.

9. Information provided to RFF by EPA, January 2001.

10. Table H-6 in Appendix H breaks down these estimates for EPA and Department of Justice.

11. Although five-year reviews may be required at non-NPL sites (sites proposed to the NPL, removal-only sites, and NPL equivalent sites), our estimates are for NPL sites only.

12. We assume that all final and deleted NPL sites will require five-year reviews during this period except for 129 "walk-away" sites identified by EPA and 3 NPL sites that had only removal actions—all with actions completed in the early 1980s. EPA gave RFF a list of these walk-away sites, where five-year reviews will not be required, in June 2000.

13. The trigger dates for five-year reviews are specified in EPA guidance. U.S. EPA, *Supplemental Five-Year Review Guidance*, July 26, 1994, p. 2.

14. The cost of these other site-specific activities was incurred at NPL equivalent and other non-NPL sites, but not removal-only sites. Expenditures for other site-specific activities at removal-only sites are included under the removal program.

Appendix F: Building Blocks of the RFF Future Cost Model

1. U.S. Environmental Protection Agency (EPA), *Groundwater Cleanup: Overview of Operating Experience at 28 Sites,* EPA 542-R-99-006, 1999.

2. U.S. EPA, Office of the Inspector General, *Superfund Backlog of Five-Year Review Reports Increased Nearly Threefold,* 1999, p. 10.

Index